# MOTION
# AND
# POWER

# MOTION AND POWER

*The continuity of every motion and the motion of every moving body depend upon the maintenance of power of the mover.*

Leonardo da Vinci

*HANS M. KOLSTEE, M.S, P.E.*

Queensborough Community College
of the City University of New York

PRENTICE-HALL, INC., Englewood Cliffs, New Jersey 07632

*Library of Congress Cataloging in Publication Data*

Kolstee, Hans M
    Motion and power.

    Bibliography: p.
    Includes index.
    1. Machinery, Kinematics of.    2. Machinery, Dynamics
of.    3. Power transmission.    I. Title.
TJ175.K66        621.8'11            80-23428
ISBN 0-13-602953-1

Editorial/production supervision by Ellen De Filippis and Theodore Pastrick
Interior design by Theodore Pastrick
Cover design by Wanda Lubelska
Manufacturing buyer: Joyce Levatino

© 1982 by Prentice-Hall, Inc., Englewood Cliffs, N.J.  07632

Printed in the United States of America

10   9   8   7   6   5   4   3   2   1

Prentice-Hall International, Inc., *London*
Prentice-Hall of Australia Pty. Limited, *Sydney*
Prentice-Hall of Canada, Ltd., *Toronto*
Prentice-Hall of India Private Limited, *New Delhi*
Prentice-Hall of Japan, Inc., *Tokyo*
Prentice-Hall of Southeast Asia Pte. Ltd., *Singapore*
Whitehall Books Limited, *Wellington, New Zealand*

# CONTENTS

# PREFACE

There is a definite need for a practical, vocationally oriented book on machine elements for training those who will work in design, drafting, technical testing, industrial management, mechanical inspection, field service, and many other technological careers. These jobs require a practical knowledge of machine elements and power transmission and of related drawings, specifications, catalogs, and industrial procedures.

This book has been developed to fill the vocational and educational needs of those studying for these careers in colleges and technical institutes. Within the time limitations of a brief, introductory course, this book presents a maximum of useful knowledge and training with a minimum of mathematical and theoretical involvement. Furthermore, this book should be useful to those now working in maintenance and technical positions in industry and to those who want a brief, comprehensive, and practical understanding of machine functions and construction.

Students of diverse educational backgrounds study machine element courses for several different reasons. Some study the subject in order to design novel and unusual machinery or to engage in research and development in this field. Others study the subject to acquire a general understanding of machine elements and power transmission needed by personnel engaged in managerial and technical industrial jobs.

This book addresses itself to the second group, and accordingly, it maintains a balance between being too complex and theoretical and too superficial and general. Information on many aspects of machine design is presented at a professionally useful level, together with data from commercial catalogs, data sheets, and other information sources used by designers in industry.

Drafting exercises in the book are representative of actual mechanical design problems related to general machinery, aerospace, marine, and other mechanical applications. In order to conserve time, the drafting exercises have been simplified, but yet the essential ideas and the professional scope have been preserved. This book includes a number of special features which enhance and broaden the training of technical students. For example, numerous photographs, illustrations, drawings, examples of industrial design problems, and commercial data sheets have been included. Brief questions at the ends of most chapters have been provided to develop the user's understanding and to review the basic ideas.

Knowledge of correct technical terminology is an important part of training. Accordingly, this terminology is used throughout the text but in a manner that does not make the material either too difficult or uninteresting. In order to give the instructor maximum flexibility in using this text, all chapters (after the first four) are practically self-contained. Thus the instructor may change the order of presentation or may omit chapters to suit his particular objectives. The last chapter, on OSHA regulations pertaining to power transmission, should be of special interest to personnel now active in related mechanical fields, and it covers an important aspect of modern mechanical training.

*Hans M. Kolstee*

# ACKNOWLEDGEMENTS

The author gratefully acknowledges the kind assistance of the following companies in making available illustrative material as well as technical and catalog data.

Ajax Flexible Coupling Co., Westfield, New York

Amcam Corporation, Farmington, Connecticut

Boston Gear, Quincy, Massachusetts

Curtis Universal Joint Co., Inc., Springfield, Massachusetts

Dana Corporation, Warren, Michigan

Dresser Industries Inc., Worthington, Ohio

Eaton Corporation, Cleveland, Ohio

Emerson Electric Co., Maysville, Kentucky

Falk Corporation, Milwaukee, Wisconsin

FMC Corporation, Philadelphia, Pennsylvania

The Gates Rubber Company, Denver, Colorado

Machine Design, Cleveland, Ohio
Macmillan Publishing Co., New York, New York
McGraw-Hill Inc., New York, New York
New Departure/Hyatt Bearings, Inc., Sandusky, Ohio
Reliance Electric Co., Inc., Mishawaka, Indiana
Renold Inc., Westfield, New York
SKF Industries, Inc., King of Prussia, Pennsylvania
Thomson Industries, Inc., Manhasset, New York
The Torrington Company, Torrington, Connecticut
Uniroyal Industrial Products, Philadelphia, Pennsylvania
U.S. Department of Labor, Occupational Safety & Health
    Administration, Washington, DC
Van Doorne Transmissie B.V., Eindhoven, The Netherlands
Winsmith Division of UMC, Springville, New York
T.B. Wood's Sons Co., Chambersburg, Pennsylvania
Zeromax Industries, Inc., Minneapolis, Minnesota

# MOTION
# AND
# POWER

# 1

# INTRODUCTION
# AND GENERAL
# PERSPECTIVE

*Drafting Instruments Needed and Precision Drafting*
*Techniques Used for the Drafting Projects.*

All machinery that has moving parts, of whatever nature, contains motion-transmitting elements. As we shall see in Chapter 4, the transmission of motion usually implies the transmission of power. A number of machine elements acting together for the transmission of motion and power is called a *mechanism*.

In an internal combustion engine the straight-line motion of the piston, called *translation*, is converted into rotation by means of a connecting rod and crankshaft. In the same engine, rotating cams impart reciprocating translating motion to the valves. The piston rod-connecting rod-crankshaft assembly and the cam-rocker-valve assembly are examples of mechanisms. In this book we shall become acquainted with several more.

In Chapter 2 we shall first define the basic forms of motion, the study of which is called *kinematics*. In Chapter 3 we shall introduce the motion-related concepts of displacement and velocity. In Chapter 4 we shall discuss the principal forces acting on power transmission elements. Such

1

forces will also be dealt with in other chapters where called for. The study of forces acting on moving bodies in general is called *dynamics*. In Chapters 5 through 13 we shall study the most important power transmission elements and their characteristics, advantages and disadvantages. Also, in Chapter 14, we shall review the requirements of safety in the practical application of these elements and their legal aspects.

This textbook has been written with the design-drafting student in mind. At the ends of most chapters there are assignments for layout exercises on the drafting board. Almost all answers to problems can be found by graphical analysis, but occasional computation will be necessary. Also, it will often be useful to verify by computation values obtained by graphical methods. A pocket calculator will be helpful in such cases.

It is assumed that students have taken at least one semester of drafting and are familiar with the basic drafting techniques.

The following drafting tools are required for the projects:

Drafting board (minimum 20 in. × 26 in.)

T-square

One 45-45-90 triangle and one 30-60-90 triangle

Engineer's scale (50 divisions per inch), preferably with metric scale

French curves (essential)

Compass with extension leg (essential)

Dividers (essential)

Protractor (essential)

Circle template

2H and 4H pencils or 2H and 4H leads for mechanical pencils

Sandpaper pack or other sharpening means

Erasers and erasing shield

Brush

Several sheets of C-size vellum

Accuracy in layout work, especially for the purpose of obtaining graphical solutions to numerical problems, is mandatory. The work should be carried out with a 4H or harder pencil, the point of which must be kept sharp at all times. For greatest accuracy, a scale larger than one (i.e., 2:1, 5:1, or even 10:1) will reduce errors proportionally.

We shall now describe some special techniques used in precision layout work.

## 1.1 TRANSFERRING DISTANCES FROM THE ENGINEER'S SCALE WITH AN ACCURACY OF .010 in.

The inch on the engineer's scale is divided into 50 equal parts of .020 in. each. Distances ending in multiples of .010 in. can be transferred from this scale with the aid of a divider point or a needle point, which points to the proper division line for even multiples of .010 in. (.020, .040, .060, etc.) and between division lines for odd multiples of .010 in. (.010, .030, .050, etc.) (See Fig. 1.1). A small circle is drawn around the needle point hole in the vellum in order to facilitate finding the hole again. The accuracy of .010 in. is easily obtainable with the naked eye, but if necessary, a magnifying glass may be used.

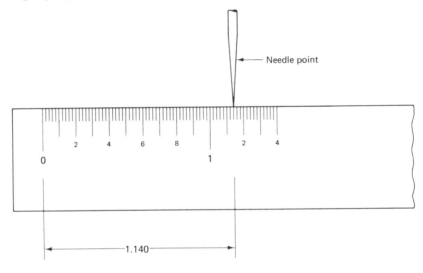

Figure 1.1

## 1.2 DIVIDING A LINE INTO A NUMBER OF EQUAL PARTS

A pair of dividers is set at the estimated length of a part and then made to "walk" along the line (see Fig. 1.2). The setting of the dividers is corrected by trial and error until the correct distance between the points is obtained.

Another method, which is also suitable for divisions of a certain (given) ratio, is shown in Fig. 1.3(a) and (b). The divisions obtained are proportional to those given due to the proportionality of the triangles 01a, 02b, etc., since lines 1a, 2b, etc., are parallel.

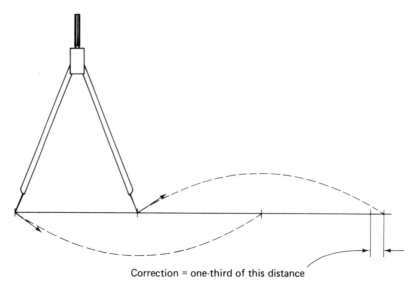

Correction = one-third of this distance

**Figure 1.2**

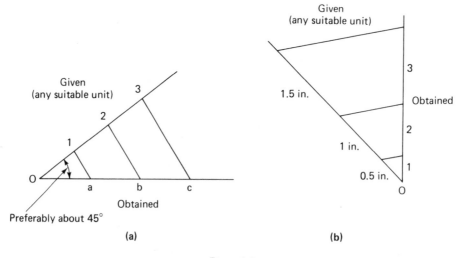

Given
(any suitable unit)

3

2

1

O

a       b       c

Obtained

Preferably about 45°

**(a)**

Given
(any suitable unit)

1.5 in.

3

Obtained

2

1 in.

1

0.5 in.

O

**(b)**

**Figure 1.3**

4

## 1.3 DRAWING IRREGULAR CURVES WITH THE AID OF A FRENCH CURVE

In the study and design of certain mechanisms, such as linkages and cams, it is often necessary to plot an irregular curve by connecting a series of points lying on that curve. This is usually accomplished with a template, the outline of which consists of several irregular, interconnected curves. Such a template is called a *French curve*. Many different outlines exist, but in most cases one or two will serve the purpose. It is almost impossible to draw an irregular curve without moving the template at least once. In most cases, parts of the curve will fit some portion of the template, thus the complete curve consists of a number of sections connected together to form a smooth line.

Figure 1.4 shows how a simple curve may be drawn by shifting the template around. Since the technique is essentially one of trial and error, several trial curves may have to be drawn and erased before a satisfactory one is obtained.

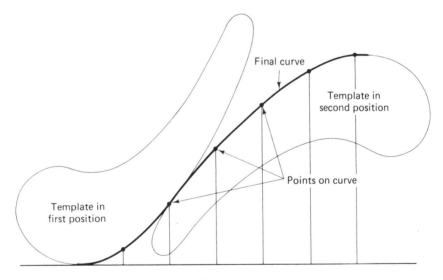

**Figure 1.4**

## 1.4 MEASURING THE LENGTH OF AN IRREGULAR CURVE WITH A DIVIDER

The length of a circular arc can be computed once the angle it subtends is known. The computation is shown in Chapter 2. There is no simple computation method to establish the length of an irregular curve. If an

5

exact answer is required, an instrument called a *curvimeter* may be used. A small wheel at one end of this instrument is made to roll along the curve from beginning to end. The distance traveled, as indicated by the number of revolutions of the wheel, can be read directly on a dial.

If an approximate value is satisfactory, a divider may be used. The points are set at a certain small distance, for example, .25 in. for small curves or .5 in. for larger curves. The divider is then made to "walk" along the curve in a manner similar to that shown in Fig. 1.5. The number of turns required for the entire distance is then counted. The number of turns plus one times the distance between the divider points, plus the remainder, if any, is the approximate length of the curve.

Since we have substituted a number of short straight lines for the curve, its actual length will always be somewhat greater than the answer obtained by this method.

The setting of the divider points should be very accurate (dividers adjusted with a fine-threaded screw are best) since any error in the setting will affect the answer by a factor equal to the number of turns plus one.

The accuracy of the divider setting may be verified as follows. Draw a straight line and transfer to it a distance equal to, say, ten times the intended divider setting, for instance 5 in. if the divider setting is .5 in. or 3.75 in. if the setting is .375 in., and so on. "Walk" the divider along the line from one end to the other. If the setting is accurate, the divider will end exactly on the mark. If it is not, the setting may be altered and another trial may be made.

## 1.5  DIVIDING A CIRCLE INTO SIX EQUAL PARTS

Although dividing a circle into six equal parts is easily accomplished with a 30-60-90° triangle, the method described below is a good exercise in the accurate use of dividers.

Set a pair of dividers as accurately as possible to the radius ($r$) of a given circle. Starting at a given point, "walk" the dividers around the circle. If this is done accurately, one should arrive exactly at the starting point after six divisions have been made.

It is also possible to make three divisions going clockwise and another three going counterclockwise. Again, the divisions should end up at the same point, that is, 180° away from the starting point.

The method is essentially one of finding the corner points of six identical, equilateral triangles of side $r$, with one common corner and two common sides each. See Fig. 1.6.

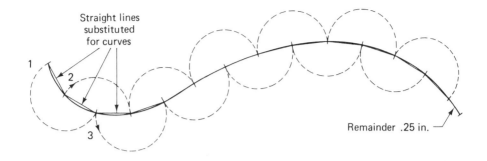

Straight lines substituted for curves

1

2

3

Remainder .25 in.

Number of turns: 12

Length of curve: $(12 + 1) \times 0.5 + 0.25 = 6.75$ in. (approximate)

**Figure 1.5**

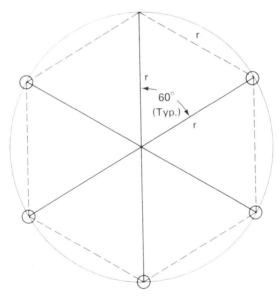

$r$

$r$

$r$

60°
(Typ.)

**Figure 1.6**

## PROBLEMS*

1. Draw four horizontal lines at the left top of the paper. Starting at a vertical line 1 in. from the left side of the paper, draw the first line 1 in. from the top. Draw the other lines in a similar manner, each 1

*Use size C vellum

7

in. lower than the previous one. Using the engineer's scale, lay out exact distances of 3.370, 5.480, 1.930, and 6.740. Use a needle point or a divider point to mark the exact locations and indicate pricked holes by drawing small circles around them (approximately 3/10 in. in diameter; use a circle template).

2. Using dividers, divide the distances laid out in Prob. 1 into two, three, four, and five equal parts, respectively.

3. Draw a circle of 2 in. radius with its center 3 in. from the top of the paper and 6 in. from the right edge. Using dividers, set at exactly 2 in., (transfer from the circle you just drew), divide the circle into six equal arcs. Indicate the locations of the pricked holes.

4. Draw a line across the bottom of the paper 1 in. away from the bottom edge. Label it the $x$-axis. Draw a vertical line 1 in. away from the left edge and label it the $y$-axis. Label the intersection 0. Divide the $x$-axis into ten equal parts, each part measuring 1 in., and number the divisions 1 to 10. Draw vertical lines approximately 10 in. long through each division. Find log 2 on your calculator (.30103), multiply it by 10, and round the answer off to three places to the right of the decimal point. (You will obtain 3.010). Mark off this number of inches on the line marked 2. Do the same with the remaining numbers, respectively lines (log 1 = 0).

   Using a French curve, connect points $y$ = 0, $x$ = 1, and the tops of all the vertical distances just transferred. You have now obtained a typical curve of the function $y$ = log $x$ for values of $x$ between 1 and 10.

5. Set your dividers at a distance of .375 and determine the approximate length of the curve you just drew. Do the same with the dividers set at a distance of .5. If you have worked accurately, your second answer should be a little smaller than your first. Why?

# 2

# MOTION

*Linear and Angular Displacement. Paths of Points on Rotating Bodies. Combined Motion. Basic Four-Bar Linkage. Mechanisms to Generate Specific Paths. Applications.*

Motion can be defined as *changing of location* with respect to a given point or object. This point or object is usually considered stationary. Motion is also possible with respect to a moving point or object, in which case it is called *relative motion*.

The result of motion is *displacement*. When a body travels along a line, *translation* or *linear motion* is taking place. The *path* or *locus* the body follows may be curved or straight. However, the *displacement* of the body is the *straight-line distance* between the point of departure and the final position. It is measured in any unit, such as in inches, centimeters, etc. (see Fig. 2.1).

When a body has undergone linear displacement, all points on that body have been displaced an equal amount. Any line or axis of the body has moved in such a manner that at any point of its path it has remained parallel to its starting position (Fig. 2.1). To determine the path followed by the translated body, any point on the body will do.

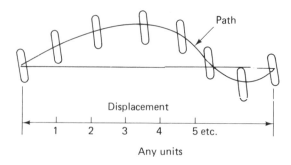

Figure 2.1

When motion occurs after which any line or axis on the body is no longer parallel to its starting position, *angular* or *rotational* motion has occurred (see Fig. 2.2). Both angular and linear displacement may take place in three dimensions, i.e., in space, but in this chapter we shall confine ourselves to motion in one plane, i.e., in two dimensions only.

When translation and rotation take place simultaneously, *combined motion is taking place*. An example is a connecting rod in an internal combustion engine.

In true rotation, or angular displacement, all points on the body rotate around a *center*. This center may be either on the body or outside of it.

Once a body has made one complete revolution, or any whole number of revolutions, all points on that body have returned to their starting positions. Once part of a revolution or a fractional number of revolutions has taken place, all points on the body except the center of rotation have been *displaced*. The amount of displacement of each point is proportional to its distance from the center of rotation (see $A_1 B_1$, $A_2 B_2$, and $A_3 B_3$ in Fig. 2.3).

Rotational or angular displacement is measured by the angle through which the body has rotated, such as $38°$, $371°$, etc. In the second example ($371°$) the displacement is the same as if the body had rotated $11°$ only. For displacement determination it is usually more convenient to indicate angular displacement after subtracting any multiples of $360°$. For example, instead of $810°$, the equivalent angle of $810° - (2 \times 360°) = 90°$ is used.

If a radius of a circle were curved and placed on the outline of its own circle in such a manner that it would coincide with a portion of that circle, it would be found to subtend (or span) an angle of $\pm 57.3°$.

Any angle may be expressed as a number of radii or as a fraction of a radius. This unit is called a *radian*. By definition, an angle of $57.3°$

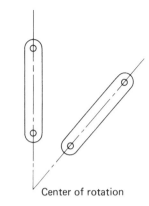

Center of rotation

Figure 2.2

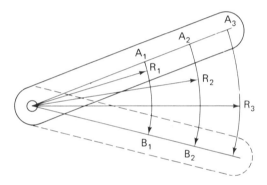

Figure 2.3

is equal to 1 radian. This method of expressing the size of an angle has certain advantages in computation, as we shall shortly see.

Since the circumference of a circle is equal to $2\pi r$, it follows that $360°$ is equal to $2\pi$ radians, $90°$ is equal to $\frac{1}{2}\pi$ radians, and so on. Any angle of $\alpha°$ is equal to $\alpha°/57.3°$ radians. (Observe that the degrees cancel out in division.)

When we call an angle expressed in radians $\theta$ (the Greek letter theta), then we have for any angle $\alpha$ in degrees

$$\theta = \frac{\alpha°}{57.3°}$$

This new unit makes it possible to establish a simple relationship between the distance traveled by any point on a body that has rotated over a given

11

angle, provided, of course, that the distance of the point to the center of rotation is known.

In Fig. 2.4 a point has traveled from $A$ to $B$ along arc $AB$. Since for a full rotation (360°), the distance traveled would have been $2\pi r$, it follows that for any angle of $\alpha°$, the distance equals

$$L = \frac{\alpha°}{360°} \times 2\pi r$$

Expressing both $\alpha°$ and $360°$ in radians, we get

$$L = \frac{\theta}{2\pi} \times 2\pi r = \theta r$$

This general equation $L = \theta r$ is valid for any angle, provided that $\theta$ is expressed in radians. It is much easier to use than the cumbersome expression

$$L = \frac{\alpha°}{360°} \times 2\pi r$$

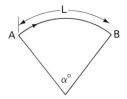

**Figure 2.4**

As mentioned before, when a body is partaking simultaneously in both linear and angular motion, it is said to be in *combined motion*. If we are interested in determining only the total linear and angular *displacement*, it is often convenient to *assume* that the body has first been in linear motion and that it has subsequently been rotated (see Fig. 2.5).

If, however, the *path* of the body is important, as, for example, in de-

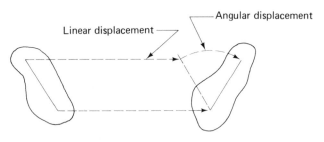

**Figure 2.5**

termining minimum internal clearances in a housing, the path of the body in combined motion must be found by drawing it, or its outline, in a series of sequential positions and plotting one or more curves of critical points (see Fig. 2.6). Such a construction can be made in a simple manner by tracing a template of the body outline, cut from cardboard or other suitable material, or by making an overlay. In the overlay method, a drawing of the moving part is placed under a piece of vellum in sequential positions. Each position is then traced on the vellum. In many cases, it is not necessary to draw the outline of the part in such detail as shown in Fig. 2.6. A *skeletonized representation* containing only critical points, such as hinge points (represented by small circles) or extremities, connected by straight lines is sufficient. Figure 2.7 shows four mechanisms in skeleton representation.

In many machines it is necessary to convert rotary motion into either rectilinear motion (i.e., motion in a straight line) or oscillating motion (i.e., reversing, cyclical rotation over a relatively small angle).

Examples of conversion mechanisms are (a) a rotating crank, connecting rod, and slider; (b) a rotating crank, connecting rod, and oscillating crank; (c) a Scotch yoke, and (d) a cam and follower (see Fig. 2.7). A well-known example of the crank and slider mechanism is found in the crankshaft, connecting rod, and piston assembly of internal combustion engines, which also have cam and follower mechanisms in the different valve actuation systems. The rotating crank, connecting rod (or *coupler*), and oscillating crank mechanism is one form of the *basic four-bar linkage*. The fourth bar is formed by the stationary support of the rotating points of the two cranks (see Fig. 2.7). The four-bar linkage is the basis of many other, more complicated linkages. By varying the length of the crank and

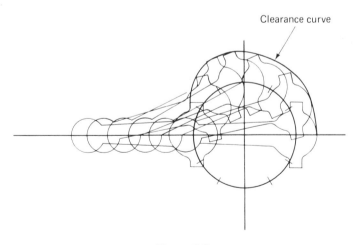

Clearance curve

Figure 2.6

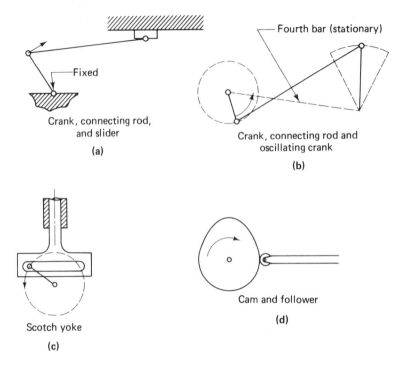

Fourth bar (stationary)

Fixed

Crank, connecting rod,
and slider

(a)

Crank, connecting rod and
oscillating crank

(b)

Scotch yoke

(c)

Cam and follower

(d)

Figure 2.7

coupler and by selecting points on the coupler, it is possible to generate innumerable looped curves, such as shown in Fig. 2.8. Motion in such paths is useful in tool and die work and in many types of automatic equipment such as packaging, wrapping, filling, and sealing machinery.

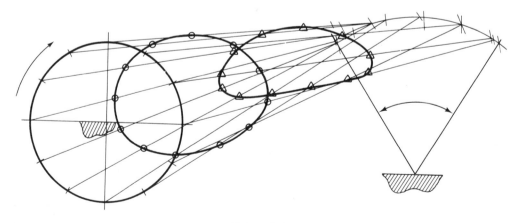

Figure 2.8

A catalog compiled by Hrones and Nelson* shows 700 different linkages and the various paths generated by different points on the coupler. The designer may select a suitable path configuration for his particular design from this catalog.

Other methods used to generate a given path are the *fixed template* and the *combined cams* methods.

The fixed template is actually a translating cam and is used for instance for tool motions in certain automatic lathes (see Fig. 2.9). A fixed template is also used in a semiautomatic milling machine called a *profiler*. In this machine a two-dimensional template is traced by a probe which causes the cutter to follow a corresponding path.

In a *shimmy die* the vertical motion of the actuating press is converted into horizontal motion by means of a fixed template.

The combined cams method can be used to generate many different curved and rectilinear (with straight lines) paths. In Fig. 2.10 two synchronized plate cams, with a 90° offset, are used to obtain an L-shaped path. For further details on cams in general, see Chapter 13.

There are other mechanisms of obtaining mathematically defined paths. We are already familiar with the rotating crank that describes a circle. It is also possible, by simple means, to describe a straight line, a

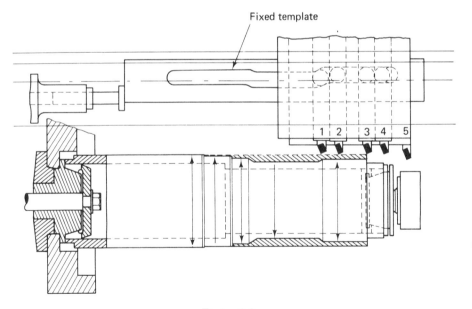

Figure 2.9

*Hrones and Nelson "Analysis of the Four Bar Linkage" (New York: John Wiley and Sons).

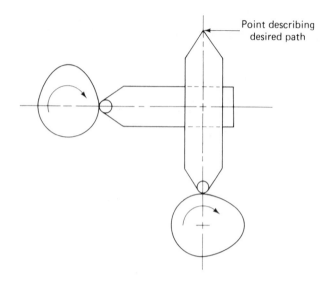

Figure 2.10

parabola, an ellipse, an involute, a cycloid, and several other curves by the mechanisms described below.

## 2.1 STRAIGHT-LINE MECHANISMS

*WATT'S STRAIGHT-LINE LINKAGE.* The Watt's Straight-line linkage is a form of four-bar linkage. Cranks *AB* and *CD* in Fig. 2.11 are of equal length and may oscillate around fixed points *A* and *C*, respectively. Point *E* lies at the center of coupler *BD*. The arrangement of parts is

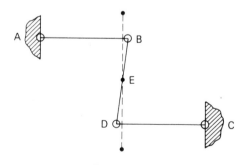

Figure 2.11

as shown. The path described by $E$ on rotation of the cranks is approximately straight for the distance shown.

*ROBERTS' STRAIGHT-LINE LINKAGE.* Cranks $AB$ and $CD$ in Fig. 2.12 are of equal length and may oscillate through small angles around $A$ and $C$. Coupler $BD$ is T-shaped and symmetrical around a vertical axis. Point $E$ at its tip lies on line $AB$ when the coupler is in its central position. Point $E$ describes an approximately straight line for some distance from its central position.

*ISOSCELES LINKAGE.* Link $AB$ in Fig. 2.13 may rotate around $A$ and is hingeably connected to the center of link $CD$ which is twice the

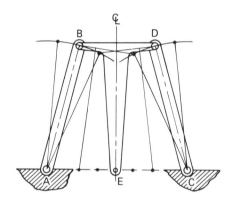

Figure 2.12

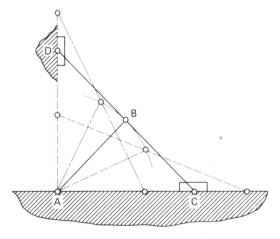

Figure 2.13

length of *AB*. Point *C* may slide in a horizontal plane. Point *D* describes an exactly straight line over its entire path.

   *EPICYCLIC STRAIGHT-LINE MECHANISM.* The stationary internal gear *A* in Fig. 2.14 has a mating pinion *B* of exactly one-half the number of teeth (i.e., it has one-half the gear's pitch diameter). Any point on the pitch circle of the pinion describes a straight diametrical line when *B* rotates inside *A*.

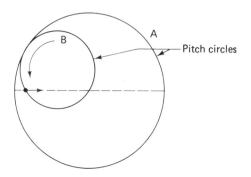

**Figure 2.14**

## 2.2   *ELLIPSE-DESCRIBING MECHANISM*

When link *CD* in Fig. 2.15 is extended beyond point *C*, any point *E* on the extension describes an ellipse when crank *AB* is rotated over 360°. Note that *C* and *D* slide along the *X* and *Y* axes, respectively.

## 2.3   *INVOLUTE CURVE-DESCRIBING MECHANISM*

A tooth rack *A* can be made to act with a mating gear *B* in the manner shown in Fig. 2.16. A point *C* on the pitch line of the rack describes an involute when, starting from a position on the gear pitch circle, the rack is rotated around the stationary gear.

## 2.4   *CYCLOID CURVE-DESCRIBING MECHANISM*

This mechanism is similar to the mechanism in Fig. 2.16, except here the path of a point on the gear instead of a point on the rack is considered. In this case, the rack is stationary, and a point on the pitch circle of the gear, traveling from tangency to tangency (1 and 2) with the pitch line of the rack, describes a cycloid (see Fig. 2.17).

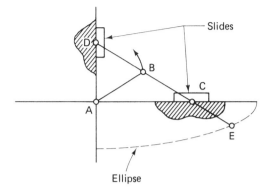

Slides

D

B

Slides

C

A

E

Ellipse

**Figure 2.15**

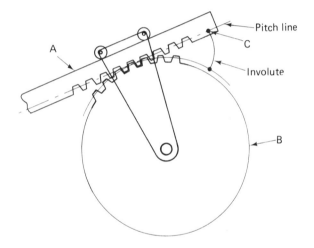

Pitch line

A

C

Involute

B

Note: Pitch circle of gear
must be tangent to
pitch line of rack

**Figure 2.16**

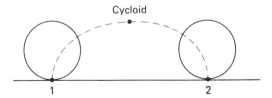

Cycloid

1

2

**Figure 2.17**

## 2.5 PANTOGRAPH

This simple instrument is a great time-saver in the drafting room where enlargement or reduction of outlines or curves is often necessary. Note in Fig. 2.18 that $BC = DE$ and $CE = BD$ and that points $A$, $E$ and $P$ lie on the same line. In use, point $A$ is held stationary.

When an outline must be reduced, point $P$ is made to trace the outline and point $E$ gives the reduced outline. When an outline must be enlarged, point $E$ is made to trace the outline and point $P$ gives the enlarged outline.

The action of the pantograph is based on the constancy of ratio $AE/AP$ for any distance $AP$ (or $AE$), which because of the similarity of triangles $ACE$ and $ABP$ is equal to $AC/AB = BD/BP$. The same ratio is also equal to the ratio of reduction. Its inverse is equal to the ratio of enlargement.

In designing linkages it may be necessary to determine the dimensions of a crank, connecting rod, or follower arm when any two of the three are given. This is often possible by construction. The following example shows how a typical construction may be performed.

Given: Center distance and locations of center of driver crank $r_C$ and follower crank $r_D$; extreme positions $AD$ and $BD$ of follower crank; and length of follower crank (see Fig. 2.19).

**Required:** Length of driver crank $r_C$ and length of connecting rod $R$.

**Solution:** Draw the two extreme positions of driver crank $r_C$ and connecting rod $R$ so that they are aligned and their centerlines both pass through the center of rotation of $r_C$. Draw $BC$ first and then draw $AC$. $BC$ now equals $R + r_C$ and $AC$ equals $R - r_C$. Since $(R + r_C) - (R - r_C) = 2r_C$, graphically subtracting $AC$ from $BC$ yields $2r_C$. Now

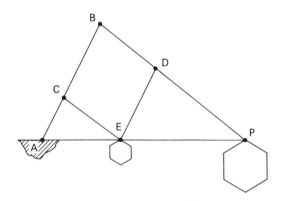

Figure 2.18

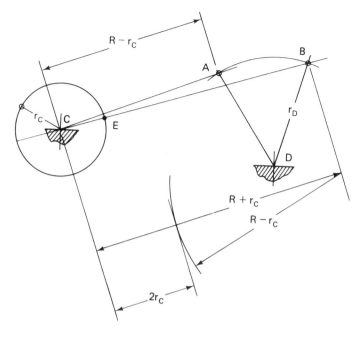

Figure 2.19

draw the crank circle. $R$ is found to be the distance from the intersection $E$ of the crank circle of $r_C$ with $BC$, to $B$.

In a similar manner, a crank, connecting rod, and slider mechanism may be constructed, as the following example will demonstrate.

Given: The distance of the center of the driver crank to the line of slider motion, the travel of the slider, and the length of the connecting rod (see Fig. 2.20).

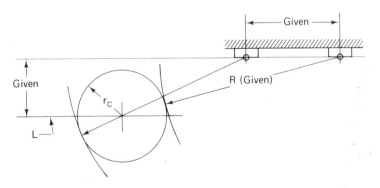

Figure 2.20

**Required:** The length of the driver crank $r_C$.

**Solution:** Draw two circular arcs of a radius equal to the length of the connecting rod so that the arcs intersect with a line L parallel to the line of slider motion and at the given distance. The centers of the arcs are the extreme positions of the slider. The crank circle of $r_C$ is now tangent to each of these arcs, and its center lies on line L. By trial and error construct the crank circle and measure the length of the crank.

## PROBLEMS

1.  How many degrees are there in an angle of 1.45 radians? How many radians are there in 132°?

2.  Define the motion (rotation, translation, or combined motion) of the following mechanisms: a bicycle chain; the rocker arms of an overhead valve engine; the front axle of a car that has independent suspension; a car's window wipers; a bus' window wipers.

3.  Refer to Fig. 2.8. On size C vellum make a layout of a similar crank and oscillating follower mechanism. Use the following dimensions: a connecting rod of length 10 in.; a follower crank of length 7 in.; and an angle of 80° (40° left and right of the vertical). The distances of the driver and follower crank centers are approximately 9 in. horizontally and 1 in. vertically. Center the layout properly. Draw a horizontal line 1 in. higher than the follower crank center. By construction (similar to that shown in Fig. 2.20) determine the path of the driver crank and thus the crank center and radius. Then do the following:

    (*a*) Starting at the top of the crank circle, draw driver crank positions on the crank circle in 15° increments. (See method on page 196 in Chapter 13.)

    (*b*) Draw the skeleton line representing the connecting rod at each of the 24 positions of the driver crank. On each of these lines locate points 1.5, 4, and 7 inches away from the driver crank pin center. Use a different shaped marker for each of these three points. For example, use a circle, a triangle, and a square.

    (*c*) After doing this for all 24 connecting rod positions, connect the corresponding points with a smooth curve.

4.  A connecting rod for a small compressor is dimensioned as shown in Fig. 2.21. Its driver crank has a length of 1.75 in. Using the overlay method, draw the outline of the connecting rod on a sheet of size A paper and construct the clearance curve required for 360° of crank

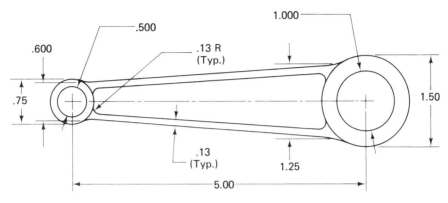

Figure 2.21

rotation. Starting with top dead center ($0°$), use $30°$ increments of crank positions.

5.  Using the following dimensions, make a layout of the Watt's linkage (Fig. 2.11): $AB = CD = 4.5$ in.; $BD = 3$ in. Draw as many different crank positions as necessary to construct the path of $E$ for all possible positions of $BD$.

6.  Using the following dimensions, make a skeletonized layout of the Roberts linkage (Fig. 2.12): $AB = CD = 5$ in.; $BD = 3.5$ in. Show the linkage in positions 2 in. and 1 in., respectively, away from center, left, and right, and show the path of $E$.

7.  An isosceles linkage (Fig. 2.13) has the following dimensions: $AB = BC = BD = 3$ in. Draw nine positions of $CD$ and show the path described by $D$.

8.  The ellipse-drawing mechanism of Fig. 2.15 has the following dimensions: $AB = BC = BD = CE = 3.75$ in. Draw three positions each in four quadrants, including the horizontal and vertical positions, so that you can construct a complete ellipse.

9.  Using strong cardboard strips 1 in. wide, construct a pantograph model dimensioned as follows: $AB = 5$ in.; $BP = 7.5$ in.; $AC = 2$ in. Use thumbtacks pointing up for the hinge points. Make a hole at $E$ and $P$ so that you can insert a pencil point. Trace a simple outline at $E$, such as a triangle, square, etc., so that point $P$ describes an enlargement of it. By measuring, verify that the enlargement ratio is $AP/AE = 2.5$.

10. In the crank, connecting rod, and slide mechanism shown in Fig. 2.20 let the connecting rod be 7.5 in., the slide travel 2.25 in., and the distance from the center of the crank to the line of slide motion (marked "Given" on the left of the drawing) 1.875 in. Determine the length of the crank.

# 3

# VELOCITY AND SPEED

*Linear and Angular Velocity, Velocities of Points on Rotating Bodies. Vector Components, Resultants, Addition, Polygon. Effective Components. Rigid Body Concept. Instant Center.*

In Chapter 2 we familiarized ourselves with the concept of motion as a change of position. Thus far we have not discussed the *rate* at which this position change takes place. This time rate is called the *velocity*.

*Linear velocity* is the linear displacement or translation per unit time. If equal distances are traveled in equal periods of time, the velocity is *constant* or *uniform*. In that case, the velocity is numerically equal to the total displacement divided by the number of time units it has taken for this displacement, or $V = S/T$, where $V$ is the velocity and $S$ is the displacement, in consistent units, and $T$ is the number of time units (seconds, minutes, hours, etc.).

When the velocity is not constant, $V = S/T$ gives the *average* or *mean* velocity. This is a theoretical number only, since most of the time the object moves at a velocity either greater or smaller than average.

In case of variable velocity, the point, line, or body under consideration is subject to *acceleration* when the velocity increases or *deceleration* when the velocity decreases. To find the velocity at any particular instant, we must use a small increment of the distance and a correspondingly small increment of the time instead of the total distance $S$ and the total time $T$. In this way, the velocity *change* becomes negligible. Since the value of the velocity found in this manner is valid for an instant only, because it is constantly changing, we now speak of the *instantaneous velocity* $V_{inst} = \Delta S/\Delta T$. ($\Delta$ = delta, a Greek letter corresponding to our $D$ and used for small increments.)

When a point travels along a curved path, the *direction* of the velocity at any position is given by the *tangent* to the path at that point. This tangent is sometimes difficult to construct for curves that have changing curvature, i.e., those that do not have an easily determined center of curvature. In such cases, the tangent can be constructed as follows. Put a small, rectangular mirror astride the curve at the point under consideration (see Fig. 3.1) and position the mirror in such a way that the curve reflected in the mirror appears to be a smooth continuation of the one on the paper. Draw a line along the base of the mirror. This line points in the direction of the center of curvature at that point, so that a line perpendicular to it and touching the curve will be the desired tangent.

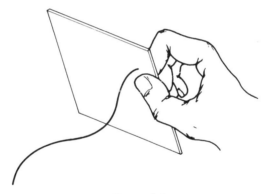

Figure 3.1

Strictly speaking, the equation $V = \Delta S/\Delta T$ is only true for straight-line motion, but it may also be used for determining $V$ in cases of curvilinear (curved-line) motion, provided $\Delta S$ is taken so small that it may be considered a straight line for all practical purposes.

When a body is in translation, *all points on that body have the same velocity*. If this were not the case, the body would deform while in motion. This rule will prove useful in solving certain problems later in this chapter.

A velocity may be represented graphically by a *vector*. A vector has the shape of an arrow and its length is equal to the numerical value of the velocity (i.e., centimeters, inches, or any other length unit). The *sense* of the velocity is indicated by the direction in which the arrow points. The *inclination* of the velocity is indicated by the angle the body of the arrow makes with the horizontal or with the $x$-axis. The sense and inclination together determine the *direction* of the velocity. Vectors are also used in other fields to represent quantities which also have magnitude and direction. Vectors are the graphical equivalent of numbers.

At this point it is appropriate to mention the difference between *speed* and *velocity*. Speed is numerically equal to the magnitude of the velocity but it does not imply direction. Velocity has both magnitude and direction.

When a point or a body is at the same time under the influence of two or more velocities, the overall effect is the same as if the point were under the influence of the *vectored sum* of these velocities. The summation of vectors produces a *resultant*. The replacement of vectors by their resultant is called the *composition* of vectors. Figure 3.2 shows the summation or composition of two vectors, $V_1$ and $V_2$. That the resultant $R$ must be the diagonal in a parallelogram with sides $V_1$ and $V_2$ is readily understood when one imagines that point $O$ first traveled to point $V_1$ in unit time and then continued toward the right over a distance equal to $O - V_2$, again in unit time.

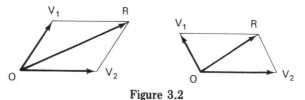

Figure 3.2

To simplify the construction needed to find a resultant, we draw the composing vectors in such a manner that the beginning of the second vector is connected to the tip of the first. We then find the resultant by drawing a line from the origin $O$ of the first vector to the tip of the second [see Fig. 3.3(a)]. In effect, only one-half the parallelogram in Fig. 3.2 is used. The resultant of more than two vectors can be found in a similar manner [see Fig. 3.3(b)].

A vector may also be considered the resultant of any two composing vectors, provided only that the latter form the sides of a parallelogram with the original vector forming the diagonal in between [see Fig. 3.3(c)]. The operation of replacing a vector by its components is called the *resolution* of a vector and is the opposite of finding a resultant.

It is clear from Fig. 3.3 that an infinite number of pairs of composing vectors have the same resultant ($V_1$ and $V_2$; $V_3$ and $V_4$; $V_5$ and $V_6$; etc.). A special group among these is formed by pairs that are perpendicular to one another, such as $V_3$ and $V_4$. We shall use such perpendicular components later in this chapter to find the instantaneous velocity of elements of linkages and other mechanisms at any particular point of their path.

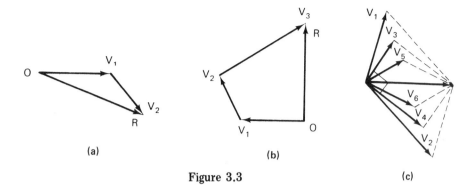

(a)

(b)

**Figure 3.3**

(c)

*Rotational or angular velocity* is the velocity of any line (a point cannot have angular velocity) on a rotating body. It describes an *angle* and is thus expressed in degrees or (more often) in *radians* per unit time. Numerically, angular velocity is found by dividing the total angular displacement $\theta$ by the time it takes for the displacement (hours, seconds, etc). In equation form it is $\omega = \theta/T$. ($\omega$ is the Greek letter omega and $\theta$ is the Greek letter theta which is used for angular displacement.) Both $\omega$ and $\theta$ are in consistent units (degrees or radians).

In similarity with $V = S/T$, $\omega = \theta/T$ gives the *average or mean* angular velocity and is thus suitable for constant or uniform velocity only.

For cases of non-uniform angular velocity, i.e., a velocity subject to acceleration or deceleration, the *instantaneous angular velocity* is given by $\omega = \Delta\theta/\Delta T$ in which $\Delta\theta$ is a small increment of angular motion and $\Delta T$ is the corresponding small increment of time in which the motion takes place. For a *point* on a body, the following relationship exists between linear and angular velocity:

$$V = \omega r \quad \text{or} \quad \omega = \frac{V}{r}$$

This can be proven as follows. We know that for linear motion, $V = \Delta S/\Delta T$ for small increments of path and time. We also know that $\omega = \Delta\theta/\Delta T$ for similar small increments. The length of the path along a circular arc is given by $S = \theta r$ [$\theta$ in radians (see Chapter 2)]. For very small arcs, we have

$$\Delta S = \Delta\theta r$$

(See Fig. 3.4.) Since $\Delta S$ is a very small arc, it may be considered a straight line. Dividing both sides of this equation by $\Delta T$, we get

$$\frac{\Delta S}{\Delta T} = \frac{\Delta\theta r}{\Delta T}$$

but $\Delta S/\Delta T = V$ (valid when $\Delta S$ is a straight line). Therefore,

$$V = \frac{\Delta\theta r}{\Delta T}$$

27

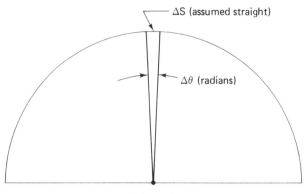

Figure 3.4

Since $\Delta\theta/\Delta T = \omega$, $V = \omega r$, or $\omega = V/r$.

The equation $V = \omega r$ can be used to prove that the magnitude of the velocities of points on a rotating body is proportional to the points' respective distances from the center of rotation. For proof, consider the body shown in Fig. 3.5. This body rotates clockwise around point $C$. Point 1 has a velocity $V_1 = \omega_1 r_1$. Point 2 has a velocity $V_2 = \omega_2 r_2$. When considered *at the same moment*, $\omega_1 = \omega_2$ even when $\omega$ is not constant.

Dividing the two velocities, we get

$$\frac{V_1}{V_2} = \frac{\omega_1 r_1}{\omega_2 r_2} \quad \text{or} \quad \frac{V_1}{V_2} = \frac{r_1}{r_2}$$

which was to be proven. This rule has important consequences in the study of gearing, belting and chain in which it is used to find speed ratios.

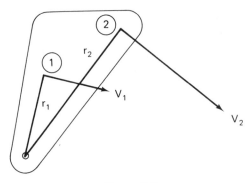

Figure 3.5

In designing machine elements it is often necessary to determine the velocity of points on links, cranks, sliders, etc., when the velocity of one element (usually the driver crank) is given. There exist several methods by

28

which this can be accomplished. In this chapter we shall use the method of *effective components*. This method is based on the fact that when a body has a velocity component in a given direction and also a velocity component perpendicular to that direction, the former is not affected by the latter. The following example will illustrate the use of this method.

**Given:** In Fig. 3.6 crank $r_1$ is 1.5 in. long and rotates clockwise at 3 revolutions per second. Crank $r_2$ is 3.5 in. long and link $AB$ is 4.25 in. long.

**Required:** The instantaneous angular velocity $\omega_2$ of crank $r_2$ in the position shown in Fig. 3.6.

**Solution:** Let the angular velocity of crank $r_1$ be $\omega_1$, expressed in radians. One revolution equals $2\pi$ rads, so that in this case the angular velocity $\omega_1 = 6\pi$ rad/s.

The linear velocity of $A$ is given by the equation $V_1 = \omega_1 r_1$. Since $r_1 = 1.5$ in., $V_1 = 6\pi \times 1.5$ in./s = 28.27 in./s. It would be impractical to draw vector $V_A$ at full scale. Assume that a vector of 2 in. long represents 28.27 in./s. We are interested in the velocity of link $AB$ since it drives crank $r_2$. If $V_A$ is projected on $AB$, $V_A$ is effectively resolved into one component perpendicular to $AB$ ($V_{AP}$) and one in the direction $AB$ ($ecV_A$). The latter is called the *effective component* of $V_A$ since it alone of the two has any effect on the motion of $AB$ *in the direction A-B*. Since $V_{AP}$ does not affect the motion of link $AB$ *in the direction A-B*, it does not affect the motion of point $B$.

As stated before, all points on a given line on a moving body must have the same velocity in the direction of that line because otherwise

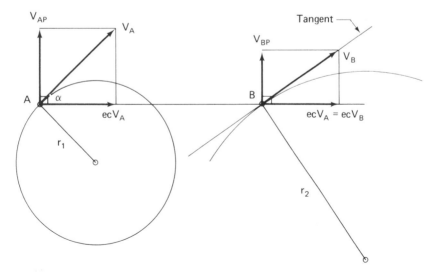

Figure 3.6

the body would deform while in motion. We may therefore transfer $ec\,V_A$ in the direction *A-B* to point *B*. This point is at the end of crank $r_2$ which, due to its *constrained motion*, can only have a velocity in the direction of the tangent to the arc of rotation of crank $r_2$ at *B*. This tangent can be drawn perpendicular to $r_2$, as shown.

Since the velocity vector of *B* must lie on this tangent line, $ec\,V_A$ must be one component of two vectors, the resultant of which is the velocity vector of *B*. The only component vector not affecting $ec\,V_A$ is one perpendicular to *AB* at point *B*.

We can now draw the parallelogram of component velocities, one of which is $ec\,V_A = ec\,V_B$, and the resultant $V_B$.

Since all vectors were drawn to the same scale as used for $V_A$, we can find the magnitude of $V_B$ by comparing it to $V_A$. Since $V_A = 2$ in. and represents 28.27 in., we can write the following equation:

$$\frac{\text{actual length of } V_B}{\text{actual length of } V_A} = \frac{\text{scale length of } V_B}{2 \text{ in.}}$$

from which:

$$\text{actual length of } V_B = \frac{\text{scale length of } V_B \times \text{actual length } V_A}{2 \text{ in.}}$$

The scale length of $V_B$ can be measured by using the dividers and the decimal inch scale (see method described in Chapter 1). The scale length of $V_B$ can also be determined by trigonometry, provided that angles $\alpha$ and $\beta$ are known. ($\alpha$ and $\beta$ are the Greek letters alpha and beta and correspond to our a and b. They are often used to denote known angles.) We then have

$$V_B = \frac{V_A \cos \alpha}{\cos \beta}$$

Using the first method of computation, we get

$$V_B = \frac{1.75 \times 28.27}{2} = 24.74 \text{ in./s}$$

It now remains for us to find the instantaneous angular velocity of $R_2$. Using the equation $\omega_2 = V_2/r_2$, we get

$$\omega_2 = \frac{24.74}{3.5} = 7.07 \text{ rad/s}$$

To facilitate the construction of and working with effective components, remember the following two rules:

1.  An effective component is always smaller than the velocity vector from which it is derived.

2. The vector of a *constrained motion* (such as the motion of a point on a crank or a slider on a flat surface) can never *be* an effective component, but it may *have* effective components.

## 3.1 INSTANT CENTER

Consider body $A$ in Fig. 3.7(a) rotating clockwise around point $O$. We can find the velocities of points $A_1$ and $A_2$ by means of the equation $V = \omega r$ when $\omega$, $r_1$, and $r_2$ are known. Now imagine body $A$ shrinking to the outline of a link $(B)$ in Fig. 3.7(b). Points $B_1$ and $B_2$ correspond to points $A_1$ and $A_2$ and have the same velocity vectors. We can still imagine that $B$ has rotated around an external center $O$. The difficulty is to find its location. If, however, we know the direction of the velocities of two points on the body, we can find the center of rotation by drawing lines through these points perpendicular to the respective velocity vectors. The intersection point of these perpendiculars is then the center of rotation [Fig. 3.7(b)].

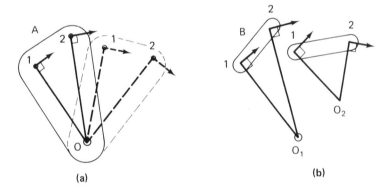

(a)                                  (b)

**Figure 3.7**

When the velocities of points $B_1$ and $B_2$ are not uniform but vary constantly in magnitude and direction, as is the case in many connecting rods, we may still use this method. The intersection of the two perpendiculars is then called the *instant*, or *instantaneous, center*. It is the point around which the body is rotating *at that particular moment and point of its path*. If we were to make a motion picture of the moving body and its velocity vectors $B_1$ and $B_2$, one frame might show the condition shown in Fig. 3.7(b) left and a later frame might show the condition shown in Fig. 3.7(b) right.

Once we have found the instant center of rotation of a body, we can determine the magnitude and direction of the velocity of *any* point on the

body by drawing a line from that point to the center, drawing a perpendicular line through the point in question, and then computing the velocity from $V_1/r_1 = V_2/r_2$. The following example shows how this method
is applied.

**Given:** The mechanism shown in Fig. 3.8(a). In the position shown,
crank *AB* has an instantaneous, counterclockwise angular velocity of
3 rad/s.

**Required:** Direction and magnitude of the instantaneous linear velocity of point *E* on link *BD*.

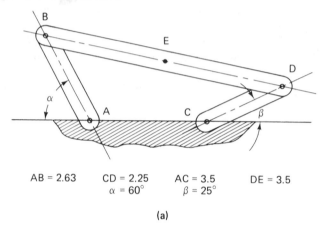

$$AB = 2.63 \qquad CD = 2.25 \qquad AC = 3.5 \qquad DE = 3.5$$
$$\alpha = 60° \qquad \beta = 25°$$

(a)

**Figure 3.8**

**Solution:** First find the magnitude and direction of two velocity
vectors on link *BD* (at *B* and *D*) by effective components [see Fig.
3.8(b)]. $V_B = \omega r = 3 \times 2.63 = 7.89$ in./s. Select a scale such that
$V_B$ is represented by a vector of 1.5 in. long. After $V_D$ has been constructed, you will find that it has a scale length of 1.93 in. The actual
linear velocity of $V_D$ is then $1.93/1.5 \times 7.89 = 10.15$ in./s.

Extending cranks *AB* and *CB* downward gives the instant center of
rotation *O*.

Draw *OE* and erect a perpendicular to it at *E* in the direction of
motion. Now find the scale value of $V_E$ from $V_E/EO = V_B/OB$ or
$V_E = EO \times V_B/OB = 3.2 \times 1.5/4.125 = 1.16$ in. (See p. 30 for
equation.) The actual value of $V_E$ is again found from

$$\frac{1.16 \times 7.89}{1.5} = 5.1 \text{ in./s } Ans.$$

The construction for the instant center used in the example above is
only practical when the angle between the two velocity vectors is not too
small, since otherwise the intersection of the perpendicular lines would
fall beyond the confines of the drafting table. In such cases, other methods

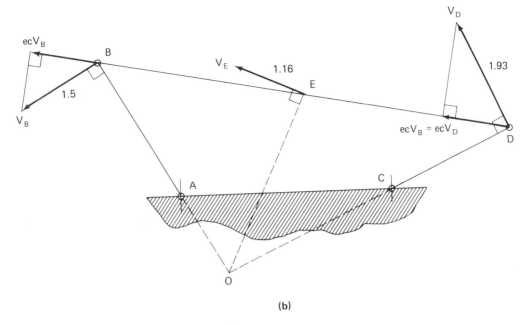

**(b)**

**Figure 3-8 cont.**

of velocity analysis may be used, but they are beyond the scope of this text.

A special case of effective components may exist in which a constrained motion vector is resolved into two other constrained motion vectors. Consider the crank and offset Scotch yoke shown in Fig. 3.9. Let the velocity vector of the tip of the crank be $V_c$. Slider $A$ can only move along line $ab$, and the stem of the Scotch yoke$^Y$ can only move along line

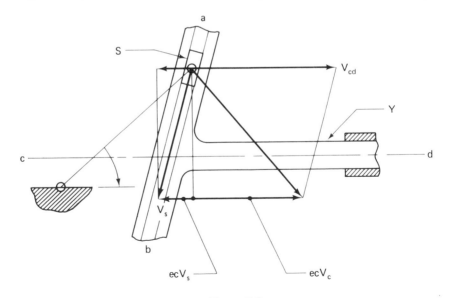

**Figure 3.9**

*cd*. Thus, the only possible composing vectors of $V_c$ can be $V_s$ and $V_{cd}$. However, these two are not mutually perpendicular. Therefore, $V_{cd}$ is not the effective component of $V_c$, although it does lie in the direction of *cd*.

$V_s$ can be resolved into a component perpendicular to *cd* and one in the direction of *cd*. This latter goes in a direction opposite to $V_y$ and thus diminishes it. The actual instantaneous velocity of $Y$ $(V_y)$ is therefore $V_{cd} - ecV_s$.

## PROBLEMS

1.  Find the resultants of the pairs of vectors shown in Fig. 3.10.

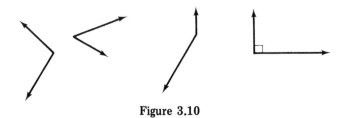

**Figure 3.10**

2.  Resolve the vectors in Fig. 3.11 into two components each in the direction indicated by the dotted lines.

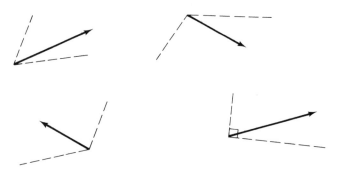

**Figure 3.11**

3.  In the crank-connecting rod-piston mechanism shown in Fig. 3.12 resolve the crankpin velocity vector shown into one velocity vector

**Figure 3.12**

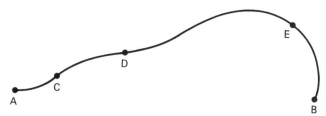

Figure 3.13

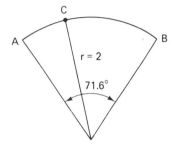

Figure 3.14

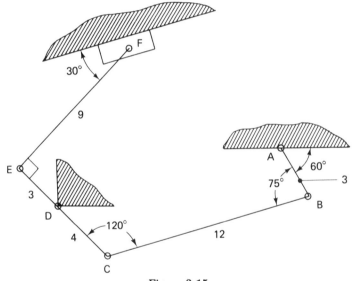

Figure 3.15

in the direction of the coupler and another one perpendicular to that direction.

4. A point travels from $A$ to $B$ along a path represented by the irregular curve shown in Fig. 3.13. It has a constant velocity of 1.5 in./s. Sketch the velocity vectors at points $C$, $D$, and $E$.

5. The radius $r$ of the circular arc $AB$ in Fig. 3.14 rotates from $A$ to $B$ in 4 s. What is the angular velocity in radians per second? What is the instantaneous velocity (in inches per second) of point $C$? Draw the vector representing this velocity.

6. In the mechanism shown in Fig. 3.15 $A$ and $D$ are fixed points. Slider

*F* moves along a fixed surface. Crank *AB* has an instantaneous, counterclockwise velocity of 1.5 rad/s.

(*a*)  Make a full-scale layout of this mechanism on size C vellum and find the instantaneous velocity of slider *F* by the construction of effective components and the application of $V_1 r_1 = V_2 r_2$ for points *C* and *E* on cranks *DC* and *DE*. Assume that vector $V_B$ is 2 in. long.

(*b*)  Verify your answer by trigonometric computation.

(*c*)  Find the instant center of link *BC* and link *EF*.

7.  Crank *C* in Fig. 3.16 rotates clockwise at a constant velocity of ½ rad/s around point *A*, driving slider *D* around in circular groove *A*. *AB* = 1.25 in. Make a full-scale skeleton layout of this mechanism (center lines only) on size B paper. Construct instantaneous velocity vectors for slider *D* for points 1, 2, 3, 4, 5, 6, and 7 as shown for

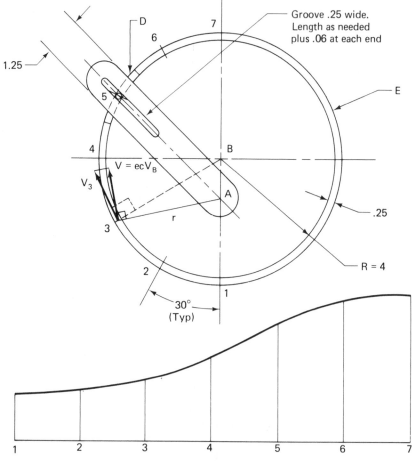

**Figure 3.16**

position 3 (not to scale). First, measure the distance from *A* to the center of the groove (*r*) and then find the velocity relative to point *A* from $V = \omega r$. Draw all vectors to their actual scale. Since slider *D* has constrained motion, it can only move in the direction of the tangent to the groove at each point. In the example at location 3, *V* must therefore be the effective component of $V_3$. By erecting a perpendicular line to the end of *V* you can draw $V_3$ as shown. After finding all velocity vectors, draw a time-velocity diagram *for the slider* as shown (not to scale) at the bottom of your sheet and demonstrate the velocity increase from point 1 to point 7. Do not scale the drawing. Measure the *x*-axis divisions (i.e., distance from point 1 to point 2 along centerline of groove) by using dividers or by computation. Use full-scale velocity vectors.

8. Crank *R* in Fig. 3.17 rotates at constant velocity in 1 s from position *A* to position *B*, while slide S, driven by connecting rod *L*, moves from position *C* to position *D* in the same time.

   (*a*)  Make a full-scale layout of this mechanism in the two positions shown. (Remember that *AC* = *BD* since connecting rod *L* does not change in length.) Locate point *O* approximately 2 in. above the bottom edge of size C vellum.

   (*b*)  Find the velocity of slide *S* at positions *C* and *D*. Draw vectors at half-scale.

9. In the mechanism in Fig. 3.18 the crank *AB* rotates counterclockwise at 2 rad/s, moving slider *C* through link *BC*. Point *D* on link *BC* at 5 in. distance from *B* is connected to crank *E* by link *DF*. Make a full-scale layout of this mechanism on size C vellum and find the angular

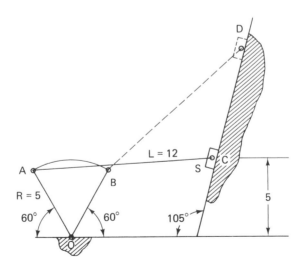

Figure 3.17

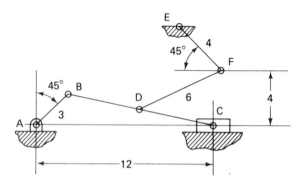

Figure 3.18

velocity of crank *EF by construction.* Show vector $V_B$ at one-third scale.

10. A crank is 5 in. long and has a linear velocity of 7.5 in./s at its tip. What is the linear velocity of a point 1.75 in. from the center of rotation?

11. Explain why all velocities on a given line and in the same direction on a translating body must be equal.

12. When a crank drives a connecting rod, at what angle between rod and crank is the instantaneous velocity of the tip of the crank equal to the instantaneous velocity of the connecting rod?

13. The crank in Prob. 12 turns at constant angular velocity. Why must we still speak of the "instantaneous velocity" of the connecting rod? (Assume that the rod drives a piston.)

14. In the piston-connecting rod-crank mechanism shown in skeleton form in Fig. 3.19 *PR* = 12 in., angle *PRQ* = 24°, and *QR* = 3.5 in. The piston velocity is 2 in./s.
    (*a*) Find the instant center of the connecting rod in the position shown and at point *S* (bottom dead center).
    (*b*) What is the velocity of point *T* on the connecting rod halfway between *P* and *Q*?
    (*c*) What is the angular velocity of the crank in the position shown?

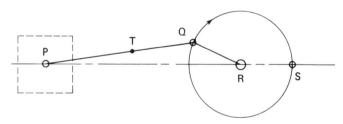

Figure 3.19

# 4

# WORK
# AND POWER

*Work and Horsepower. Torque. Efficiency of Power*
*Transmission. Service Factors.*

*Work* is defined as the product of a *force* (pounds, newtons, or other
unit) and the distance through which this force acts (feet, meters, or other
unit).

Let us take a pair of coordinates $x$ and $y$ and designate the $x$-axis
as measuring the distance and the $y$-axis as measuring the force. We can
now represent the amount of work performed by a given combination
of force and distance by an area, bounded by the $x$-and $y$-axes, a horizontal
line through the value of the force on the $y$-axis, and a vertical line through
the value of the distance on the $x$-axis (see Fig. 4.1.).

Area $A$ in Fig. 4.1 represents the work performed by a force of 200
lb acting over a distance of 2 ft and is numerically equal to 400 ft-lb.
Area $B$ represents the work performed by a force of 50 lb acting over a
distance of 8 ft and is also equal to 400 ft-lb.

It will be apparent from these two cases that there are any number of
possible combinations of force and distance the product of which is 400
ft-lb. It may be thought that area $A$ represents the work performed by

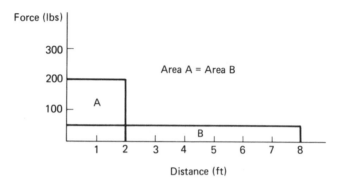

Figure 4.1

a horse dragging a load that has a horizontal resistance of 200 lb over a distance of 2 ft and that area *B* represents the work performed by a person dragging a load that has a horizontal resistance of 50 lb over a distance of 8 ft. The *amount of work* performed by the horse and the person is the same in this case.

Since a horse is obviously much stronger than a person, it is necessary to introduce another factor in the work concept to indicate the different rates of performing work. This factor is *time*. If it is assumed that both the horse and the person drag their loads at the same velocity, it will be clear that the horse could perform four times as much work as the person *in the same span of time.*

The customary unit used to express the *ability to perform work*, or the work absorbed by driven equipment, is the *horsepower*. Horsepower is numerically equal to 550 ft-lb per second. The word horsepower is somewhat of a misnomer, since the average working performance of a horse is approximately equal to only one-half of 1 hp.

Using this unit makes it possible to make meaningful comparisons of the *maximum possible* output of *prime movers* such as electric motors, internal combustion engines, turbines, etc., and similarly the power *requirements* of various types of driven equipment such as pumps, machine tools, elevators, packaging machines, and so on. The word "maximum" in the preceding sentence means that prime movers only develop as much horsepower as is demanded of them by the driven equipment. For example, the designation "a 5-hp three-phase motor" means that this motor can perform work at *any rate up to 5 hp*. The average automobile engine develops most of the time only a fraction of its rated (maximum) horsepower capacity.

The horsepower unit may also be defined as 746 W. This information is useful in determining the size of wiring, switch gear, breakers, and other electrical equipment required when electric motors are used as prime movers.

In the first sentence of this chapter we defined work as the product of a force and the distance through which it acts. In the examples in Fig. 4.1 we assumed that the distance was a straight line. Let us now consider the case in which a person turns a crank, say to lift a bucket of water from a well, by means of a rope wound around a drum connected to a crank. The force exerted by the person's arm now follows a *circular path*. The work performed per revolution of the crank is equal to

$$\frac{2\pi rF}{12} \text{ ft-lb}$$

where $r$ is the length of the crank in inches and $F$ is the force exerted by the person. If the crank makes $n$ revolutions per minute, the work *per second* is

$$\frac{2\pi rFn}{12 \times 60} \text{ ft-lb}$$

Since, as mentioned before, a work rate of 550 ft-lbs/sec is called 1 hp, we can now state that the work performance is:

$$\text{hp} = \frac{2\pi rFn}{12 \times 60 \times 550}$$

The product of the force and the length of the crank $F \times r$ is called the *torque* and is usually expressed (as in this case) in *inch-pounds*.

When the horsepower and the revolutions per minute are known, the torque $T$ may be found by rearranging the above equation as follows:

$$T = Fr = \frac{12 \times 60 \times 550 \times \text{hp}}{2\pi n} \text{ (in.-lb)}$$

Multiplying, dividing, and substituting revolutions per minute (rpm) for $n$, we get

$$T = \frac{63,000 \times \text{hp}}{\text{rpm}} \text{ (approximately)}$$

This is its customary form in industry.

Note that in this equation the torque $T$ is proportional to the transmitted horsepower but inversely proportional to the revolutions per minute. The latter relationship is of vital importance in the design of gears, chain, belts, couplings, and shafting.

Within the limitations imposed by centrifugal force, lubrication, and some other factors not discussed further here, for a given horsepower load, the faster the shaft runs, the smaller the forces on gear teeth, chain, belts, etc., and therefore the lighter they can be.

It is also clear that for a given value of $T = Fr$, when $r$ is doubled, $F$ is halved, and so on. This in turn means that for a given torque, the larger the diameter ($2r$) of the pulley, sprocket, gear, etc., the lighter the belt or chain or the smaller the tooth size.

The following examples illustrate how the above equations are used.

**Given:** A pump absorbs 7.5 hp when running at 600 rpm.
**Required:** The torque in the shaft.
**Solution:**

$$T = \frac{63,000 \times \text{hp}}{\text{rpm}}$$

$$= \frac{63,000 \times 7.5}{600} = 788 \text{ in.-lb}$$

**Given:** A person lifts a bucket of water that has a total weight of 26 lb over a distance of 14 ft.
**Required:** (a) The work performed (ignore friction effects). (b) The length of the crank if the force exerted by the person is not to exceed 8 lb and if the diameter of the drum is 6 in.
**Solution:**

($a$) Work performed = force times distance
$$= 26 \times 14 = 364 \text{ ft-lb} \quad \text{Ans.}$$
($b$) Torque = $F \times r = 26 \times 3$ ($r = \frac{1}{2}$ diameter) = 78 in.-lb.
Then $78 = 8 \times r_1$ and $r_1 = 78/8 = 9.75$ in.  Ans.
($r$ = drum radius and $r_1$ = length of crank.)

## 4.2  EFFICIENCY

There are no mechanical power transmission or conversion (for example, linear to rotational, or vice versa) processes known in which *all* of the energy (horsepower) is transmitted or converted. Some (usually small) portion of the energy is lost to friction and is changed into heat, which is subsequently dissipated to the surroundings by radiation, convection, or conduction.

Friction losses occur in bearings, belts (both externally by slipping and internally by bending), chain, gears, flexible couplings, etc. Bent shafts have internal friction losses.

The ratio of the horsepower actually transmitted (output) and the horsepower supplied by the prime mover (input) is called the *efficiency* of the drive or transmission and is often expressed as a percentage. Typical percentages will be mentioned in the chapters dealing with specific transmission elements.

A transmission system consisting of several individual units connected in series, for example, an electric motor driving a gear reducer via a V-belt, has an overall (system) efficiency equal to the *product* of the efficiencies of the individual units. If in this example the belt drive has an efficiency of 97% and the gear reducer has an efficiency of 98%, the overall efficiency of the drive is .97 $\times$ .98 = .95, or 95%.

## 4.3 SERVICE FACTORS

Since input horsepower times efficiency equals output horsepower, the theoretical input horsepower for a mechanical transmission may be determined by taking the total average horsepower consumption of the driven equipment (output horsepower) and dividing this value by the overall efficiency of the transmission.

However, many types of driven equipment have horsepower requirements that are constantly varying, sometimes with peak values ("shock loads") considerably above average. Such variations are often difficult to determine exactly. Also, starting of a prime mover may impose a momentary peak load on the transmission, which depends on the nature of the driven equipment as well as on the prime mover. For instance, a heavy flywheel or pulley causes a greater starting load than does a light fan wheel. A steam turbine creates less of a starting peak load than a steam engine does. Similar differences exist among different types of electric motors.

To ensure that a transmission system is of adequate capacity to take any additional, temporary overloads in its stride, a *service factor* is applied to the output horsepower found by the formula in the first paragraph of this section. This factor is based on previous experience with similar equipment and prime movers and in reality reflects our ignorance of actually occurring peak loads in the system. Service factors also depend on the nature of the transmission element (gear, belt, chain, etc.).

Service factors are used for determining the needed capacity of belts, gears, chain, couplings, shafting, bearings, and any other transmission elements. They can be selected from tables such as Table 4.1 to meet most of the possible combinations of driver and driven equipment. The service factor is multiplied by the theoretical horsepower mentioned above. The product, called *design horsepower*, is then used as the basis for selecting the prime mover and transmission elements. Table 4.1 is for belting only. A similar table for gearing, which has somewhat higher values, is given in Chapter 6.

TABLE 4.1
Service Factors for V-belts.

| DriveN Machine | DriveR | | | | | |
|---|---|---|---|---|---|---|
| | AC Motors: Normal Torque, Squirrel Cage, Synchronous, Split Phase.<br><br>DC Motors: Shunt Wound.<br><br>Engines: Multiple Cylinder Internal Combustion.* | | | AC Motors: High Torque, High Slip, Repulsion-Induction, Single Phase, Series Wound, Slip Ring.<br>DC Motors: Series Wound, Compound Wound.<br>Engine: Single Cylinder Internal Combustion.*<br>Line shafts    Clutches | | |
| The machines listed below are representative samples only. Select the group listed below whose load characteristics most closely approximate those of the machine being considered. | Intermittent Service | Normal Service | Continuous Service | Intermittent Service | Normal Service | Continuous Service |
| | 3-5 Hours Daily or Seasonal | 8-10 Hours Daily | 16-24 Hours Daily | 3-5 Hours Daily or Seasonal | 8-10 Hours Daily | 16-24 Hours Daily |
| Agitators for Liquids<br>Blowers and Exhausters<br>Centrifugal Pumps & Compressors<br>Fans up to 10 Horsepower<br>Light Duty Conveyors | 1.0 | 1.1 | 1.2 | 1.1 | 1.2 | 1.3 |
| Belt Conveyors For Sand, Grain, Etc.<br>Dough Mixers<br>Fans Over 10 Horsepower<br>Generators<br>Line Shafts<br>Laundry Machinery<br>Machine Tools | 1.1 | 1.2 | 1.3 | 1.2 | 1.3 | 1.4 |

TABLE 4.1 (*cont.*)
Service Factors for V-belts.

| DriveN Machine | DriveR | | | | |
|---|---|---|---|---|---|
| Punches-Presses-Shears | | | | | |
| Printing Machinery | | | | | |
| Positive Displacement Rotary Pumps | | | | | |
| Revolving and Vibrating Screens | | | | | |
| Brick Machinery | | | | | |
| Bucket Elevators | | | | | |
| Exciters | | | | | |
| Piston Compressors | | | | | |
| Conveyors (Drag-Pan-Screw) | | | | | |
| Hammer Mills | 1.2 | 1.3 | 1.4 | 1.5 | 1.6 |
| Paper Mill Beaters | | | | | |
| Piston Pumps | | | | | |
| Positive Displacement Blowers | | | | | |
| Pulverizers | | | | | |
| Saw Mill and Woodworking Machinery | | | | | |
| Textile Machinery | | | | | |
| Crushers (Jaw-Roll) | | | | | |
| Mills (Ball) | | | | | |
| Hoists | 1.3 | 1.4 | 1.5 | 1.6 | 1.8 |
| Rubber Calander-Extruders-Mills | | | | | |

\* Apply indicated service factor to continuous engine rating. Deduct 0.2 (with a minimum service factor of 1.0) when applying to maximum intermittent rating.

The use of a service factor of 2.0 is recommended for equipment subject to choking.
For Grain Milling and Elevator Equipment, see Mill Mutual Bulletin No. VB-601-62.
For Oil Field Machinery, see API specification for Oil Field V-Belting, API Standard 1B.

(Courtesy Gates Rubber Co., Denver, Colorado.)

Below is a numerical example of the use of efficiency values and service factors.

Given: A laundry machine is driven by a slip ring AC motor via a V-belt for 8 h a day. The transmission used has an overall efficiency of 94%. The machine absorbs an average of 8 hp.

Required: The minimum rating of the belt.

Solution:

$$\text{input hp} = \frac{\text{output hp}}{\text{efficiency}}$$

$$= \frac{8}{.94} = 8.5 \text{ hp (approx.)}$$

Now go to Table 4.1. Find "Laundry Machinery" in the column headed "DriveN Machine." In the right column under the heading "DriveR" find "AC motors, Slip Ring." In the middle subcolumn under the heading "Normal Service, 8-10 Hours Daily", find the service factor 1.3.

The design horsepower of the belt then equals 1.3 × 8.5 = 11 hp (approx.)

## PROBLEMS

1. A small steam turbine runs at 12,000 rpm and exerts a torque of 28 in.-lb. What horsepower does it develop?

2. A tugboat tows a barge at a speed of 7 knots. The tension in the tow-line is 12,500 lb. What is the net horsepower developed by the tug-boat? (1 knot = 6,080 ft/h.)

3. A kneading machine is driven by a belt and chain transmission. The belt drive has an efficiency of 95% and the chain drive has an efficiency of 97%. The machine absorbs 4 hp (including service factor). What should the minimum rated output of the driver be?

4. A transmission system develops an output torque of 14,700 in.-lb while running at 120 rpm. The input is 30 hp. What is the efficiency of this system?

5. An electric motor has an output of 1.75 hp at the shaft and draws 1,450 kW. What is the overall efficiency of this motor?

6. A wood planer is used an average of 4 hr/day, and needs 5 hp. It is driven by a series-wound DC motor via a timing (toothed) belt which

has an efficiency of 98%. Using Table 4.1, determine the minimum design horsepower for the belt.

7. It has been possible to determine that the highest peak torque in a certain driven machine is 74,000 in.-lb while running at 400 rpm. If the service factor for this machine and its driver is 1.5, what is the average horsepower requirement?

# 5

# SPUR GEARING

*Terminology. Involute Conjugate Curves. Law of Gearing.
Standard Spur Gear Tooth Dimensions. Commercial
Catalogs. Size Determination. Using Commercial Catalogs.*

In order to understand the action of gear teeth, it is helpful to start by considering a friction drive. A friction drive is created by the action of two cylinders rolling together, as shown schematically in Fig. 5.1. The tangential velocity $V$ at the point of contact of the two cylinders is obviously the same for each. From Chapter 3 we know that $V = \omega r$ ($\omega$ in radians per second). Then

$$\omega_1 r_1 = \omega_2 r_2$$

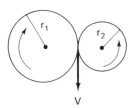

Figure 5.1

Dividing both sides by $r_1 \omega_2$, we get

$$\frac{\omega_1}{\omega_2} = \frac{r_2}{r_1}$$

in which $\omega_1 / \omega_2$ is called the *speed ratio*.

When $\omega$ is divided by $2\pi$ and multiplied by 60 it becomes revolutions per minute. Therefore,

$$\frac{\text{rpm}_1}{\text{rpm}_2} = \frac{r_2}{r_1} = \frac{d_2}{d_1}$$

or in words: The speed ratio of two cylinders in true rolling contact is equal to the *inverse* ratio of their radii or diameters.

The drawback inherent in a friction drive such as described above is that it is likely to slip when power of any consequence is transmitted. It is useful only for very small torque applications such as phonograph turntable drives and the like.

The positive prevention of slippage in the transmission of larger torque values requires the use of *teeth*. These teeth penetrate into the surface of each cylinder of the friction drive (see Fig. 5.2).

Mating cylinders or wheels provided with teeth are called *gears*. The diameter of each of the original rolling cylinders is called the *pitch diameter* and the cylinder's sectional outline is called the *pitch circle*.

The shape of the tooth outline curve must be such that *no changes in speed ratio* occur during the passing contact of each tooth with its mating tooth on the other gear. This is a basic requirement of all gearing. Curves that satisfy this requirement are called *conjugate curves*.

Several types of conjugate curves exist, but the one almost universally used today is the *involute* (see Fig. 5.3). This curve is described by a point on a string as it is being unwound from a cylinder.

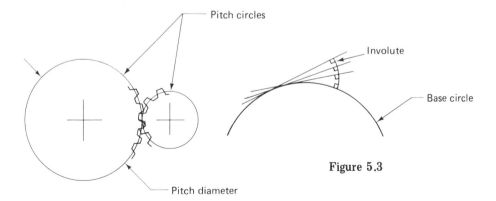

Pitch circles

Involute

Base circle

Figure 5.3

Pitch diameter

Figure 5.2

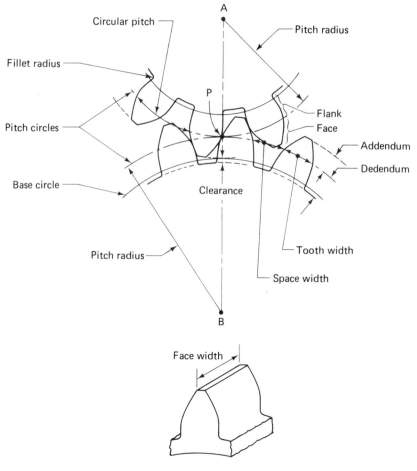

**Figure 5.4**

In *spur gears* the teeth are straight lengthwise and parallel to the axis of rotation. Spur gears are the simplest of all gears. Their nomenclature and definitions (see also Fig. 5.4) which follow are mostly applicable to all other types of gearing as well.

## 5.1  NOMENCLATURE AND DEFINITIONS

**Addendum** — The radial distance a tooth protrudes outside the pitch circle.

**Backlash** — The space width minus the tooth width.

**Base circle** — That from which the involute is generated (see Fig. 5.3).

**Base pitch** — Similar to the circular pitch (see below) measured along the base circle instead of along the pitch circle. It is numerically equal to the circular pitch times the cosine of the pressure angle.

**Center distance** — Equals the sum of the pitch *radii* or equals one-half the sum of the pitch diameters $(D_{p1} + D_{p2})/2$.

**Circular pitch** $(P_c)$ — The distance between corresponding sides of adjacent teeth; *measured along the pitch circle.*

**Clearance** — The difference between the addendum and the dedendum in mating gears.
Both backlash and clearance are essential to prevent binding of the teeth of two mating gears as a result of small variations in dimension caused by manufacturing tolerances.

**Dedendum** — The radial distance from the bottom of the groove (the root) between adjacent teeth to the pitch circle.

**Diametral pitch** $(P_d)$ — The number of teeth of a gear for each inch of pitch diameter. The circular pitch and the diametral pitch (see Fig. 5.12 for various tooth series and diametral tooth sizes) are related as follows: $P_d \times P_c = \pi$.

**Face of a tooth** — The active surface of the tooth *outside* the pitch cylinder.

**Face width** — The dimension of a tooth measured parallel to the axis of the gear.

**Fillet** — The rounded corner at the base of the tooth.

**Flank of a tooth** — The active surface of the tooth *inside* the pitch cylinder.

**Line of centers** — The line connecting the centers of two mating gears.

**Pitch diameter** $(P_d)$ — The diameter of the pitch circle. It is equal to twice the pitch *radius.*

**Pitch point** $(P)$ — The point of tangency of the pitch circles.

**Pressure angle** — The angle between the common tangent to the base circles (i.e., the line of action in Fig. 5.7) and the common tangent to the pitch circles at the pitch point. At the present time the preferred pressure angle for spur gears is $20^\circ$. In newer designs this angle replaces the value of $14.5^\circ$ formerly in general use.

**Space width** — The distance between facing sides of adjacent teeth; *measured along the pitch circle.*

**Tooth width** — The width of the tooth; *measured along the pitch circle.*

**Working depth** — The sum of the addendum of a gear and the addendum of its mating gear (i.e., twice the addendum in standard gears).

In order to mate properly, gears running together must have (1) the same pitch, (2) the same pressure angle, and (3) the same addendum and dedendum (standard gears only).

Since the number of teeth of each of two mating gears is proportional to the pitch diameter of each gear, it is easier, and thus customary, to express speed ratio $(m_\omega)$ in tooth numbers $(N)$ instead of in pitch diameters. Thus,

$$m_\omega = \frac{\text{rpm}_1}{\text{rpm}_2} = \frac{N_2}{N_1} = \frac{d_2}{d_1}$$

We have already stated that the involute is the curve almost universally used today for shaping the outline of gear teeth. Let us now analyze the action of a pair of mating involute teeth. Let $A$ and $B$ in Fig. 5.5 be the base circles of mating involute gears. Line $CD$ is a common tangent, and $AB$ is the line of centers. Assume that $CD$ is part of a string being unwound from $A$ and wound upon $B$ and that $A$ and $B$ rotate together in such a manner that string $CD$ remains taut at all times. When $A$ and $B$ start to rotate, point $C$ will leave circle $A$ and move toward circle $B$, thus describing an involute with respect to circle $A$. At the same time, however, we can imagine that, with respect to circle $B$, point $C$ traces an involute back to its origin somewhere on $B$. Adding these two involutes to the sketch, we now have the condition shown in Fig. 5.6.

It can be proven that the basic requirement for proper gear action, namely, no changes in speed ratio during the passage of any tooth, is fulfilled when the normal to the mating tooth curves passes through the pitch point at all times. Let us investigate whether this condition is satisfied by the involute.

It is clear that, since the string is taut at all times, the path of the point of contact between the two involutes is a straight line. This line intersects the line of centers ($AB$) at $P$ (see Fig. 5.7). Also, the involute is, by definition, normal to its generating line (i.e., the string) at all times (see Fig. 5.3). If we can prove that point $P$ is the pitch point, we have satisfied the above requirement.

Triangles $ACP$ and $BPD$ are similar because their respective angles are equal. As with the friction drive at the beginning of this chapter, $A$ and $B$ have the same circumferential velocity. Therefore, we may state that $AC/BD = \text{rpm}_B/\text{rpm}_A$. But, $AC/BD = AP/BP$ and $\text{rpm}_B/\text{rpm}_A$ is the speed ratio. Therefore, $AP$ and $BP$ must be the pitch radii and point $P$ must be the pitch point.

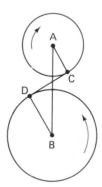

Figure 5.5

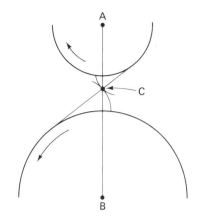

Figure 5.6

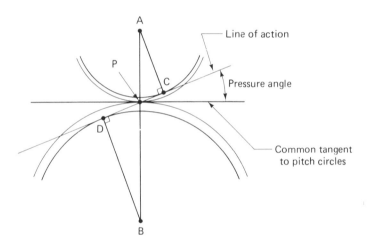

Figure 5.7

## 5.2 PATH OF CONTACT AND CONTACT RATIO

The *path of contact* is the line described by the point of contact between two mating teeth. In involute gearing the path of contact lies on the line of action. It begins where the addendum circle of the dri*ven* gear intersects the line of action and it ends where the addendum circle of the dri*ving* gear intersects the line of action. (This definition ignores possible interference conditions with pinions of small tooth numbers, which are outside the scope of this text.)

   Figure 5.8 shows that *increasing* the pressure angle will decrease the length of the path of contact (pressure angle $\alpha$ is smaller than pressure angle $\beta$; path of contact $A_1B_1$ is longer than path of contact $A_2B_2$).

   The *contact ratio* indicates the *average* number of teeth in contact for a given pair of mating gears. It is found by dividing the length of the path of contact by the base pitch. For continuous action, the contact ratio should be at least 1. In actual practice, however, it is recommended that the contact ratio be at least 1.4.

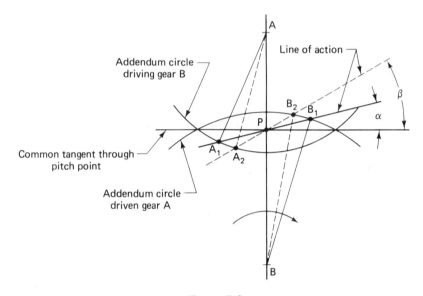

Figure 5.8

## 5.3 *DIMENSIONS OF SPUR GEARS*

Since the diametral pitch $(P_d)$ determines the physical size of a tooth (although not the shape; see below), it is logical to base the main tooth dimensions on it.

The following relations are valid for both $14\frac{1}{2}°$- and $20°$-pressure angle *standard* involute gears:

$$\text{Addendum} = \frac{1}{P_d} \qquad \begin{array}{l}\text{Clearance}\\ \text{and}\\ \text{fillet radius}\end{array} = \frac{.157}{P_d}$$

$$\text{Dedendum} = \frac{1.157}{P_d} \qquad \text{Tooth width} = \frac{P_c}{2}$$

## 5.4  STRENGTH OF INVOLUTE TEETH

The side of an involute rack tooth, i.e., a tooth of a gear having an infinitely large diameter, is straight (see Fig. 5.9). For a given pitch, the smaller the number of teeth, the more undercut the flank becomes (see Fig. 5.10).

Involute rack

**Figure 5.9**

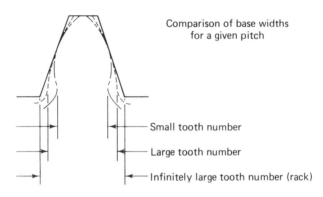

Comparison of base widths
for a given pitch

Small tooth number

Large tooth number

Infinitely large tooth number (rack)

**Figure 5.10**

It can be proven mathematically that the greatest (tensile) stress in a loaded gear tooth occurs near the base of the tooth, and it is here that the greatest number of tooth breakages takes place. It follows from the above that for a given pitch, the smaller the number of teeth, the weaker each tooth. Generally, if the gears are of the same width and material, the smaller of two mating gears will thus be the weaker and should therefore be used as the basis for size and pitch selection.

In Fig. 5.7 the base circle is closer to the pitch point in the pinion than in the gear. Since dedendum, respectively addendum, must be the same for two standard mating gears, it will be clear that part of the *gear* teeth projects inside the base circle of the *pinion* during engagement. Since the involute curve extends outside the base circle, the flank of the pinion teeth inside the base circle is no longer of involute shape.

When the teeth are *generated*, i.e., cut while rotating with the cutting tool, proper action of the teeth is ensured since any interfering portions of the pinion teeth are cut off. The drawback is that the pinion teeth become excessively undercut and thus weakened in pinions that have a very small number of teeth. For this reason, pinions having fewer than 18 teeth for the 20°-pressure angle system and fewer than 32 for the 14.5°-pressure angle system are not recommended.

## 5.5   ADVANTAGES AND DISADVANTAGES OF INVOLUTE GEARING

Involute gears have the following advantages:

1.  Since the involute rack is straight-sided, certain generating tools used to manufacture involute gears have flat faces that are easy to manufacture and to sharpen.
2.  Involute gears may operate at a center distance somewhat greater than theoretically required and still maintain proper conjugate action.
3.  The involute shape produces *generally* strong teeth.

Involute gears have the following disadvantages:

1.  As already mentioned, in pinions having a small number of teeth, the flank of the tooth becomes progressively more undercut when the number of teeth is reduced. This often weakens the tooth to an unacceptable degree.
2.  Under certain conditions not further specified here, teeth of mating gears and pinions may interfere with one another and may require modifications in the tooth outline. One modification uses a reduced addendum for the gear and an increased addendum for the pinion. This system is called *long and short addendum teeth*. Another, similar modification is called the *20°-stub tooth system*. Both modifications correct the interference conditions but at the expense of interchangeability. (See *American Gear Manufacturers' Handbook* for details on this and many other aspects of gear design.)

## 5.6   DRAWING INVOLUTE TOOTH OUTLINES

For the infrequent case that a draftsman would be called upon to draw actual tooth outlines, either of the following methods may be used to avoid the elaborate construction of involutes:

1. Using a template for the most common standard pitches and tooth numbers.
2. Approximating the involute by two circular arcs, as shown by Fig. 5.11 and Table 5.1. This method, which was originally devised by Grant, usually gives a fair approximation of the actual outline. First the circle sections are laid out, then the radial line, and finally the root fillet and both arcs of $R$ and $r$.

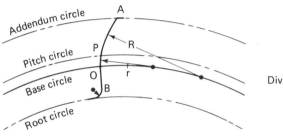

Divide radii given in Table 5.1 by diametral pitch

Figure 5.11

TABLE 5.1

| No. of Teeth | 14½° | | 20° | | No. of Teeth | 14½° | | 20° | |
|---|---|---|---|---|---|---|---|---|---|
| N | R | r | R | r | N | R | r | R | r |
| 12 | 2.87 | 0.79 | 3.21 | 1.31 | 29 | 5.24 | 2.52 | 6.25 | 3.93 |
| 13 | 3.02 | 0.88 | 3.40 | 1.45 | 30 | 5.37 | 2.63 | 6.43 | 4.10 |
| 14 | 3.17 | 0.97 | 3.58 | 1.60 | 31 | 5.51 | 2.74 | 6.60 | 4.26 |
| 15 | 3.31 | 1.06 | 3.76 | 1.75 | 32 | 5.64 | 2.85 | 6.78 | 4.42 |
| 16 | 3.46 | 1.16 | 3.94 | 1.90 | 33 | 5.77 | 2.96 | 6.95 | 4.58 |
| 17 | 3.60 | 1.26 | 4.12 | 2.05 | 34 | 5.90 | 3.07 | 7.13 | 4.74 |
| 18 | 3.74 | 1.36 | 4.30 | 2.20 | 35 | 6.03 | 3.18 | 7.30 | 4.91 |
| 19 | 3.88 | 1.46 | 4.48 | 2.35 | 36 | 6.17 | 3.29 | 7.47 | 5.07 |
| 20 | 4.02 | 1.56 | 4.66 | 2.51 | 37-39 | 6.36 | 3.46 | 7.82 | 5.32 |
| 21 | 4.16 | 1.66 | 4.84 | 2.66 | 40-44 | 6.82 | 3.86 | 8.52 | 5.90 |
| 22 | 4.29 | 1.77 | 5.02 | 2.82 | 45-50 | 7.50 | 4.46 | 9.48 | 6.76 |
| 23 | 4.43 | 1.87 | 5.20 | 2.98 | 51-60 | 8.40 | 5.28 | 10.84 | 7.92 |
| 24 | 4.57 | 1.98 | 5.37 | 3.14 | 61-72 | 9.76 | 6.54 | 12.76 | 9.68 |
| 25 | 4.70 | 2.08 | 5.55 | 3.29 | 73-90 | 11.42 | 8.14 | 15.32 | 11.96 |
| 26 | 4.84 | 2.19 | 5.73 | 3.45 | 91-120 | 0.118N | | 0.156N | |
| 27 | 4.97 | 2.30 | 5.90 | 3.61 | 121-180 | 0.122N | | 0.165N | |
| 28 | 5.11 | 2.41 | 6.08 | 3.77 | Over 180 | 0.125N | | 0.171N | |

(Courtesy: *Technical Drawing*, Giesecke, Mitchell, Spencer, and Hill; New York, Macmillan.)

## 5.7 SELECTING SPUR GEARS FROM CATALOGS

The basis for gear selection should be the pinion since it is the weaker of two mating gears of the same material and width. In some cases, it is advantageous to make the pinion from a stronger material than the gear, for example, a steel pinion mating with a cast-iron gear. This provides more equal tooth strength and often improves the wear characteristics of the teeth as well.

The procedure is as follows. The number of horsepower to be transmitted is multiplied by the service factor whenever applicable. The horsepower value found, together with pinion revolutions per minute is used to select a pitch value and tooth number for the pinion. Then a suitable gear

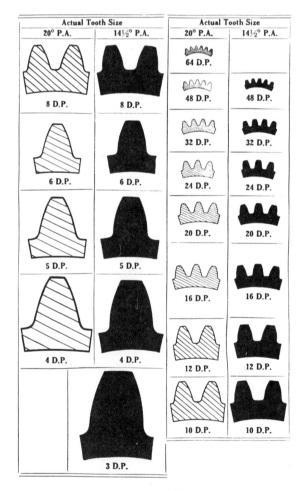

Figure 5.12
Boston Gear, Quincy, Mass

is selected and its horsepower rating is checked, because although the gear may have more teeth, it may be made from a weaker material than that of the pinion.

Since only a limited number of possible ratios can be obtained from a given range of standard gears, a compromise must often be made. If a pinion that has more teeth than originally planned for were required for another pinion-gear combination, the same pitch value could always be used since, for a given pitch, the greater the number of teeth, the stronger each tooth and the greater the horsepower capability of the pinion (see Fig. 5.10). However, when selecting pinions that have more than 20 teeth, one should verify whether the pitch line velocity recommended by the manufacturer (in this case, for a 20°-pressure angle, metallic spur gear: 1,500 rpm) will be exceeded. Find the pitch line velocity from

$$\text{pitch line velocity} = \frac{\text{rpm} \times \text{pitch diameter} \times \pi}{12} \text{ fpm}$$

An often-occurring restraint is the permissible center distance range. In such cases, the required pinion pitch diameter range must first be computed and then pinion(s) within that range selected.

Whenever possible, whole number values for the speed ratio, such as 3, 4, 7, etc., should be avoided. With such speed ratios, the same pinion tooth mates with the same teeth of the gear at all times, which may lead to uneven wear.

In specially cut gears, an extra tooth is often provided in the gear to eliminate this condition. This tooth is called a *hunting tooth*.

Below is an example of how to select a pair of spur gears from the Boston Gear Catalog.

Find a pair of spur gears to transmit 1.5 hp (service factor applied) at 1,200 driver rpm and 500 driven rpm. Center distance to be between 5 and 6; pressure angle to be 20°.

1. Since the center distance is a restraint, we must find the required pinion pitch diameter first.

$$\frac{D_p \text{ gear}}{D_p \text{ pinion}} = \frac{\text{rpm pinion}}{\text{rpm gear}}$$

$$\frac{D_{pg}}{D_{pp}} = \frac{1{,}200}{500} = 2.4 \quad \text{or} \quad D_{pg} = 2.4 D_{pp} \tag{1}$$

also:

$$\frac{D_{pg} + D_{pp}}{2} = \text{center distance } C \tag{2}$$

**20°** PRESSURE ANGLE
*(Will not operate with 14½° spurs)*

| Pitch | Teeth | 25 | 50 | 100 | 200 | 300 | 600 | 900 | 1200 | 1800 |
|---|---|---|---|---|---|---|---|---|---|---|
| | | | | | | RPM | | | | |
| **20** | 12 | – | .01 | .03 | .07 | .10 | .18 | .25 | .32 | .43 |
| **STEEL** | 16 | .01 | .03 | .05 | .10 | .15 | .28 | .38 | .48 | .62 |
| | 20 | .02 | .04 | .07 | .14 | .20 | .36 | .49 | .60 | .77 |
| | 50 | .06 | .12 | .21 | .40 | .55 | .88 | 1.10 | 1.26 | 1.48 |
| **20** | 100 | .10 | .20 | .36 | .61 | .80 | 1.15 | 1.34 | 1.40 | – |
| **IRON** | 200 | .21 | .38 | .64 | .98 | 1.19 | 1.53 | – | – | – |
| **16** | 12 | .02 | .04 | .08 | .15 | .22 | .41 | .57 | .71 | .93 |
| **STEEL** | 16 | .03 | .06 | .13 | .24 | .35 | .63 | .85 | 1.04 | 1.33 |
| | 20 | .04 | .09 | .17 | .32 | .46 | .81 | 1.08 | 1.30 | 1.62 |
| | 48 | .13 | .25 | .48 | .86 | 1.17 | 1.82 | 2.24 | 2.53 | 2.91 |
| **16** | 96 | .23 | .44 | .79 | 1.30 | 1.66 | 2.30 | 2.65 | – | – |
| **IRON** | 192 | .46 | .83 | 1.35 | 2.01 | 2.41 | – | – | – | – |
| **12** | 12 | .05 | .10 | .19 | .36 | .52 | .92 | 1.26 | 1.53 | 1.96 |
| **STEEL** | 16 | .08 | .15 | .29 | .56 | .80 | 1.39 | 1.84 | 2.21 | 2.75 |
| | 20 | .10 | .20 | .39 | .74 | 1.04 | 1.17 | 2.30 | 2.72 | 3.30 |
| | 36 | .22 | .42 | .80 | 1.39 | 1.95 | 3.04 | 3.74 | 4.22 | 4.85 |
| **12** | 84 | .47 | .88 | 1.55 | 2.53 | 3.18 | 4.30 | – | – | – |
| **IRON** | 240 | 1.27 | 2.16 | 3.29 | 4.49 | 5.11 | – | – | – | – |
| **10** | 12 | .09 | .17 | .33 | .63 | .91 | 1.60 | 2.14 | 2.58 | 3.24 |
| **STEEL** | 16 | .14 | .27 | .53 | .99 | 1.40 | 2.38 | 3.11 | 3.67 | 4.49 |
| | 20 | .19 | .37 | .70 | 1.30 | 1.81 | 3.00 | 3.84 | 4.47 | 5.34 |
| | 30 | .31 | .60 | 1.13 | 2.03 | 2.76 | 4.30 | 5.28 | 5.97 | 6.86 |
| **10** | 80 | .80 | 1.48 | 2.56 | 4.08 | 5.08 | 6.71 | – | – | – |
| **IRON** | 200 | 1.89 | 3.19 | 4.30 | 6.69 | 7.61 | – | – | – | – |
| **8** | 12 | .16 | .32 | .61 | 1.16 | 1.65 | 2.83 | 3.72 | 4.41 | 5.42 |
| **STEEL** | 16 | .26 | .51 | .97 | 1.79 | 2.51 | 4.15 | 5.31 | 6.18 | 7.38 |
| | 20 | .35 | .68 | 1.29 | 2.35 | 3.23 | 5.18 | 6.49 | 7.43 | 8.68 |
| | 32 | .62 | 1.20 | 2.16 | 3.76 | 4.99 | 7.43 | 8.88 | 10.83 | – |
| **8** | 72 | 1.31 | 2.41 | 4.14 | 6.46 | 7.94 | 10.3 | – | – | – |
| **IRON** | 192 | 3.19 | 5.45 | 8.06 | 10.7 | – | – | – | – | – |
| **6** | 12 | .38 | .75 | 1.43 | 2.65 | 3.70 | 6.12 | 7.84 | 9.12 | 10.9 |
| **STEEL** | 16 | .61 | 1.18 | 2.24 | 4.06 | 5.56 | 8.83 | 11.0 | 12.5 | 14.5 |
| | 21 | .87 | 1.69 | 3.15 | 5.56 | 7.46 | 11.37 | 13.76 | 15.38 | – |
| | 24 | 1.02 | 1.96 | 3.63 | 6.33 | 8.41 | 12.51 | 14.94 | 16.55 | – |
| **6** | 66 | 2.78 | 5.01 | 8.39 | 12.6 | 15.3 | – | – | – | – |
| **IRON** | 144 | 5.73 | 9.49 | 14.1 | 18.8 | – | – | – | – | – |
| **5** | 12 | .68 | 1.33 | 2.53 | 4.63 | 6.39 | 10.3 | 13.0 | 14.9 | 17.5 |
| **STEEL** | 16 | 1.09 | 2.10 | 3.95 | 7.03 | 9.51 | 14.7 | 17.9 | 20.2 | 23.0 |
| | 20 | 1.49 | 2.86 | 5.19 | 9.04 | 12.0 | 17.9 | 21.4 | 23.7 | – |
| **5** | 24 | 3.55 | 6.26 | 5.29 | 9.03 | 11.8 | 17.0 | 20.0 | 21.9 | – |
| **IRON** | 60 | 4.43 | 7.99 | 13.1 | 19.6 | 23.4 | – | – | – | – |
| | 180 | 11.7 | 18.2 | 25.6 | – | – | –. | – | – | – |

**Figure 5.13**
Boston Gear Quincy, Mass.

Substituting (1) into (2), we have

$$2.4D_{pp} + D_{pp} = 2C$$

When $C = 5$,

$$D_{pp} = \frac{10}{3.4} = 2.941$$

and when $C = 6$,

$$D_{pp} = \frac{12}{3.4} = 3.529$$

2. In the approximate horsepower table in Fig. 5.13, trace down column 1 (the rpm column) until you come to 1,200 rpm and find the following:

| Pitch | Teeth | hp |
|-------|-------|-----|
| STEEL | | |
| 16 | 48 | 2.53 |

then $D_p$ = 48/16 = 3 (acceptable)

This pinion would need a 2.4 × 48 = 115 teeth gear (approximately). Now go to the 20°-spur gear table for 16 pitch and find cast-iron gears of 96 and 128 teeth. Go back to the horsepower table and check the horsepower. For the 96 teeth gear, the horsepower is 2.30 for 600 rpm. It is not shown for the 128 teeth gear, which indicates that this may be too high a speed for this particular gear. The 96 teeth gear meets neither the speed ratio nor the center distance requirements, since the speed ratio is 96/48 = 2 (2.4 is needed) and

$$C = \frac{D_{pg} + D_{pp}}{2} = \frac{6 + 3}{2} = 4.5 \qquad \text{(5 is minimum)}$$

3. Continuing across the 1,200 rpm column in Fig. 5.13, we find the following:

| Pitch | Teeth | hp |
|-------|-------|-----|
| STEEL | | |
| 12 | 36 | 4.22 |

then $D_{pp}$ = 36/12 = 3 (acceptable)

The required number of teeth of a gear is 2.4 × 36 = 86 (approximately). In the spur gear table for 12 pitch (Fig. 5.14), an 86 teeth cast-iron gear is shown to be available. Go back to the horsepower table and find that at 600 rpm the gear's rating is 4.30 hp, which is more than adequate. Since the speed ratio is almost and the center distance is exactly as specified (verify!), we shall assume that they are acceptable. Although one may object that this pair of gears is somewhat overdesigned for the job, the economy and ease of using stock gears would normally outweigh this minor drawback.

## PROBLEMS

1. Construct an involute as follows: Place a size A sheet of vellum on the drafting board and fasten it with tape. Draw a vertical centerline. Draw a 3-in. radius circle with its center on the centerline approximately 4 in. from the bottom of the sheet. Starting approximately 45° to the right of the centerline (top one-half of the circle), locate, mark, and number 16 points on this circle. Going counterclockwise, make them *exactly* .25 in. apart. Use dividers. Using the right angle of a triangle, draw a tangent to the circle at each of these 16 points. The tangent lines should favor the right side of the paper. Number these

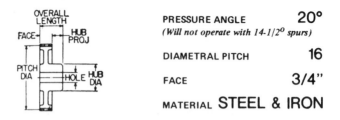

PRESSURE ANGLE **20°**
*(Will not operate with 14-1/2° spurs)*

DIAMETRAL PITCH **16**

FACE **3/4"**

MATERIAL **STEEL & IRON**

ORDER BY CATALOG OR ITEM NUMBER

| No. of Teeth | Pitch Dia. | Bore | Hub Dia. | Hub Proj. | Without Keyway & Setscrew | | With Keyway & Setscrew* | |
|---|---|---|---|---|---|---|---|---|
| | | | | | Cat. No. | Item No. | Cat. No. | Item No. |
| **STEEL** | | | | | | | | |
| 12 | .750" | 3/8" | 9/16" | 1/2" | YB12 | 09916 | YB12–3/8 | 46141 |
| 14 | .875 | 3/8 | 11/16 | 1/2 | YB14 | 09918 | YB14–3/8 | 46142 |
| 15 | .938 | 3/8 | 3/4 | 1/2 | – | – | YB15–3/8 | 45991 |
| | | 1/2 | | | YB15 | 09920 | YB15–1/2 | 46143 |
| 16 | 1.000 | 1/2 | 13/16 | 1/2 | YB16 | 09922 | YB16–1/2 | 46144 |
| 18 | 1.125 | 1/2 | 15/16 | 1/2 | YB18 | 09924 | YB18–1/2 | 46145 |
| 20 | 1.250 | 5/8 | 1-3/64 | 1/2 | YB20 | 09926 | YB20–5/8 | 46146 |
| 24 | 1.500 | 5/8 | 1-3/16 | 1/2 | YB24 | 09928 | YB24–5/8 | 46147 |
| | | 3/4 | | | – | – | YB24–3/4 | 46148 |
| 28 | 1.750 | 5/8 | 1-7/16 | 1/2 | YB28 | 09930 | YB28–5/8 | 46149 |
| | | 3/4 | | | – | – | YB28–3/4 | 46150 |
| 30 | 1.875 | 5/8 | 1-9/16 | 1/2 | YB30 | 09932 | YB30–5/8 | 46151 |
| | | 3/4 | | | – | – | YB30–3/4 | 46152 |
| | | 7/8 | | | – | – | YB30–7/8 | 46153 |
| 32 | 2.000 | 5/8 | 1-11/16 | 1/2 | YB32 | 09934 | YB32–5/8 | 46154 |
| | | 3/4 | | | – | – | YB32–3/4 | 46155 |
| | | 7/8 | | | – | – | YB32–7/8 | 46156 |
| | | 1 | | | – | – | YB32–1 | 46157 |
| 36 | 2.250 | 5/8 | 1-15/16 | 1/2 | YB36 | 09936 | – | – |
| 40 | 2.500 | 5/8 | 2-3/16 | 5/8 | YB40 | 09938 | – | – |
| 48 | 3.000 | 5/8 | 2 | 5/8 | YB48A | 10572 | – | – |
| 56 | 3.500 | 5/8 | 2-1/2 | 5/8 | YB56A | 10574 | – | – |
| 60 | 3.750 | 5/8 | 2-3/4 | 5/8 | YB60A | 10576 | – | – |
| 64 | 4.000 | 3/4 | 2-7/8 | 3/4 | YB64A | 10578 | – | – |
| 72 | 4.500 | 3/4 | 3-3/8 | 3/4 | YB72A | 10580 | – | – |
| 80 | 5.000 | 3/4 | 3-7/8 | 3/4 | YB80A | 10582 | – | – |
| **CAST IRON** | | | | | | | | |
| 96 | 6.000 | 3/4 | 1-3/4 | 3/4 | YB96 | 10584 | – | – |
| 128 | 8.000 | 3/4 | 2 | 3/4 | YB128 | 10588 | – | – |
| 144 | 9.000 | 3/4 | 2 | 3/4 | YB144 | 10590 | – | – |
| 160 | 10.000 | 7/8 | 2 | 3/4 | YB160 | 10592 | – | – |
| 192 | 12.000 | 7/8 | 2 | 1 | YB192 | 10594 | – | – |

*3/8" Bore and larger has one Setscrew.
5/8" Bore and larger has Standard Keyway at 90° to Setscrew.

**12 thru 80 Teeth – PLAIN STYLE**
**96 thru 192 Teeth – SPOKE STYLE**

**Figure 5.14(a)**
Boston Gear Quincy, Mass.

lines the same as the points to which they are tangent. Beginning with line 2, mark off .25 in. to the right of the tangency point. Use dividers left at the same setting as before. Then continuing with line 3, mark off 2 × .25 in. = .5 in.; with line 4, mark off 3 × .25 in. = .75 in., and so on to line 16. Connect the points on all these tangents by making a smooth curve (use a French curve). You now have an *approximate* involute. Why is it an approximate involute?

**20°**    PRESSURE ANGLE
*(Will not operate with 14-1/2° spurs)*

**12**      DIAMETRAL PITCH

**1"**        FACE

# STEEL & IRON MATERIAL

### ORDER BY CATALOG OR ITEM NUMBER

| No. of Teeth | Pitch Dia. | Bore | Hub Dia. | Hub Proj. | Without Keyway & Setscrew | | With Keyway & Setscrew* | |
|---|---|---|---|---|---|---|---|---|
| | | | | | Cat. No. | Item No. | Cat. No. | Item No. |
| **STEEL** | | | | | | | | |
| 12 | 1.000" | 1/2" | 3/4" | 5/8" | YD12 | 09940 | YD12–1/2 | 46158 |
| 13 | 1.083 | 5/8 | 53/64 | 5/8 | YD13 | 09942 | YD13–5/8 | 46159 |
| 14 | 1.167 | 5/8 | 29/32 | 5/8 | YD14 | 09944 | YD14–5/8 | 46160 |
| 15 | 1.250 | 5/8 | 63/64 | 5/8 | YD15 | 09946 | YD15–5/8 | 46161 |
| 16 | 1.333 | 5/8 | 1-1/16 | 5/8 | YD16 | 09948 | YD16–5/8 | 46162 |
| 18 | 1.500 | 3/4 | 1-15/64 | 5/8 | YD18 | 09950 | YD18–3/4 | 46163 |
| 20 | 1.667 | 3/4 | 1-5/16 | 5/8 | YD20 | 09952 | YD20–3/4 | 46164 |
| 21 | 1.750 | 3/4 | 1-25/64 | 5/8 | YD21 | 09954 | YD21–3/4 | 46165 |
| | | 7/8 | | | – | – | YD21–7/8 | 46166 |
| 24 | 2.000 | 3/4 | 1-41/64 | 5/8 | YD24 | 09956 | YD24–3/4 | 46167 |
| | | 7/8 | | | – | – | YD24–7/8 | 46168 |
| | | 1 | | | – | – | YD24–1 | 46169 |
| 28 | 2.333 | 3/4 | 1-63/64 | 5/8 | YD28 | 09958 | YD28–3/4 | 46170 |
| | | 7/8 | | | – | – | YD28–7/8 | 46171 |
| | | 1 | | | – | – | YD28–1 | 46172 |
| 30 | 2.500 | 3/4 | 29/64 | 5/8 | YD30 | 09960 | – | – |
| 36 | 3.000 | 3/4 | 1-15/16 | 7/8 | YD36A | 10596 | – | – |
| 42 | 3.500 | 3/4 | 2-7/16 | 7/8 | YD42A | 10598 | – | – |
| 48 | 4.000 | 7/8 | 2-7/8 | 7/8 | YD48A | 10600 | – | – |
| 54 | 4.500 | 7/8 | 3-3/8 | 7/8 | YD54A | 10602 | – | – |
| **CAST IRON** | | | | | | | | |
| 60 | 5.000 | 7/8 | 2-1/8 | 7/8 | YD60 | 10604 | – | – |
| 66 | 5.500 | 7/8 | 2-1/8 | 7/8 | YD66 | 10606 | – | – |
| 72 | 6.000 | 7/8 | 2-1/8 | 7/8 | YD72 | 10608 | – | – |
| 84 | 7.000 | 7/8 | 2-1/8 | 7/8 | YD84 | 10610 | – | – |
| 96 | 8.000 | 7/8 | 2-1/8 | 7/8 | YD96 | 10612 | – | – |
| 108 | 9.000 | 7/8 | 2-1/4 | 7/8 | YD108 | 10614 | – | – |
| 120 | 10.000 | 1 | 2-1/4 | 7/8 | YD120 | 10616 | – | – |
| 132 | 11.000 | 1 | 2-1/2 | 1 | YD132 | 10618 | – | – |
| 144 | 12.000 | 1 | 2-1/2 | 1 | YD144 | 10620 | – | – |
| 168 | 14.000 | 1 | 2-1/2 | 1 | YD168 | 10622 | – | – |
| 192 | 16.000 | 1 | 2-1/2 | 1 | YD192 | 10624 | – | – |
| 216 | 18.000 | 1 | 2-3/4 | 1 | YD216 | 10626 | – | – |

*1/2" Bore and larger has one Setscrew.
5/8" Bore and larger has Standard Keyway
at 90° from Setscrew.

**12 thru 54 Teeth — PLAIN STYLE**
**60 Teeth — WEB STYLE**
**72 thru 216 Teeth — SPOKE STYLE**

**Figure 5.14(b)**
Boston Gear Quincy, Mass.

2. Generate an involute tooth profile as follows: On a size A sheet of vellum draw (do not trace) the generating rack shown in Fig. 5.15. Note that the numbered divisions are *exactly* .25 in. apart. For greatest accuracy, transfer all these points from your scale without moving it. On a second sheet of size A vellum draw a centerline in the short direction. Draw a circular arc of about 120° with a 6.665-in. radius (use decimal scale) with its center on the centerline approximately .5

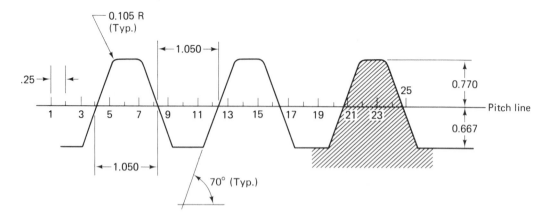

Generating rack for 1.5 $P_d$ 20° full-depth gear.
Draw on a size A vellum.

**Figure 5.15**

in. from the top of the paper. This arc represents part of the pitch circle of a 20 tooth, $1.5P_d$, 20°-pressure angle spur gear.

Following the procedure outlined in Prob. 1, locate, mark, and number 24 points on this arc *exactly* .25 in. apart. Use dividers. Draw a radius through each of these points. Now fasten the first sheet (with the rack) to your board with tape. Place the second sheet on top of the first sheet in such a manner that point 1 on the rack pitch line coincides with point 1 on the pitch circle and that the radius coincides with the vertical line through point 1 on the rack. Temporarily tape the second sheet to the board and trace the outline of the rack very accurately. Remove the tape and relocate the top sheet so that you are able to repeat the procedure for points 2 and so on (see Fig. 5.16). You should now have an outline of the tooth profile for a few teeth. The process simulates the generation of a gear by the gear shaping process, and it demonstrates the remarkable fact that a curved flank is generated by a straight-sided cutting tool. *Note:* Success with this construction depends a great deal on the accuracy of the layout of the .25 in. distances on both sheets.

3.  A 24 tooth pinion runs at 1,150 rpm and drives a 54 tooth gear. What is the speed ratio? What is the speed of the gear?

4.  A belt conveyor is driven by a shunt-wound DC motor through a pair of 20°-pressure angle spur gears. The conveyor is in operation for 8 h each day and absorbs 7.5 hp. The motor runs at 600 rpm and the conveyor runs at 180 rpm.

    (*a*) Compute the minimum required horsepower capacity of the motor.

    (*b*) Select a suitable pitch for the gears.

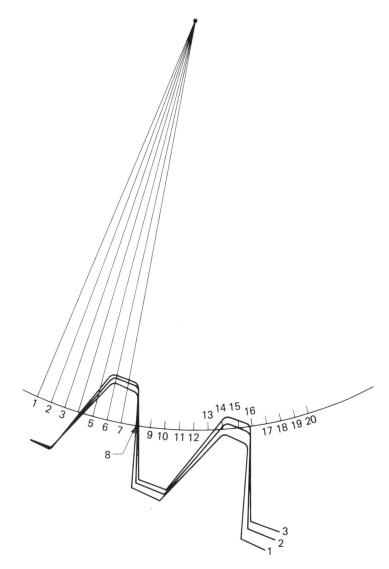

Figure 5.16

5. Using the method of approximation described on page 7, on a sheet of size A vellum make a layout of a $1.5P_d$ tooth. The gear has 20 teeth and a 20°-pressure angle. The pitch circle radius is 6.665. The addendum is .667 and the dedendum is .778. The base circle radius is 6.330. This is the same tooth as constructed in Prob. 2. Make an overlay and compare the outlines to see how accurate Grant's method is.

6. On size C vellum make a layout of two mating spur gears. Specifications are as follows: Full depth, $8P_d$; 20°-pressure angle; pinion, 24 teeth; gear, 49 teeth.

(*a*)  Start drawing a horizontal line in the middle of the paper. At 6 in. from the left edge, draw a mark on this line. Let this mark represent the center of the pinion.

(*b*)  Compute the center distance and mark the center of the gear on the same line.

(*c*)  Draw the pitch circles solid and the addendum and dedendum circles dotted.

(*d*)  Sketch in a few teeth in pinion and gear (not at the pitch point). Find the size in Fig. 5.12.

(*e*)  When the gear rotates at 650 rpm, what is the rpm of the pinion?

(*f*)  If the specifications call for a clearance of .13 all around, what is the minimum inside length of the housing?

# OTHER GEAR TYPES

*Worm Gears. Overall Speed Ratio of Gear Trains. Selecting Standard Gear Reducers from Commercial Catalogs. Five Gear Reducer types.*

The gears described below have been developed either to improve upon spur gears or to meet specific purposes.

## 6.1 HELICAL GEARS

Helical gears have teeth in the shape of a slow helix (see Fig. 6.1). Since the teeth are engaged and disengaged in a more gradual manner, helical gears run more quietly than spur gears. Because of the geometry of tooth contact, they also have greater load capacity, and more teeth are engaged at any one time.

A drawback is the axial (sideways) thrust caused by the helix angle, but this can be overcome by having the teeth of one-half of the face width cut in the shape of an opposite helix. Such gears are called *herring-bone gears.* They are more expensive to manufacture and are mostly used in heavy-duty applications such as marine reduction gears and steel mill drives.

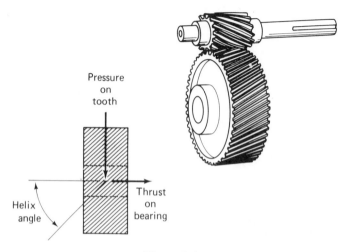

Pressure
on
tooth

Helix
angle

Thrust
on
bearing

Figure 6.1

Mating helical gears that have parallel shafts have *equal but opposite* helix angles. Mating helical gears that have identical, 45°-helix angles run at right angles to one another and are called *crossed helical gears* (Fig. 6.2).

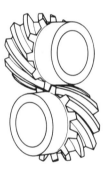

Figure 6.2

Morse Chain Division of Borg-Warner.

## 6.2  BEVEL AND MITER GEARS AND ANGULAR GEARS

These gears are used for motion and power transmission between intersecting shafts at right angles. Miter gears have a 1/1 speed ratio; bevel gears may have any ratio but ordinarily not over 6/1. Bevel and miter gears may have straight (Fig. 6.3) or spiral (Fig. 6.4) teeth and are mostly sold as matched pairs. In *angular* gears the axes meet at an angle other than 90°. For proper tooth action, the pitch cones of bevel, miter, and angular gears must have a common tangent and center (Fig. 6.5).

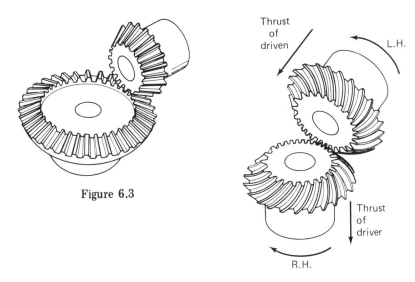

Figure 6.3

Figure 6.4

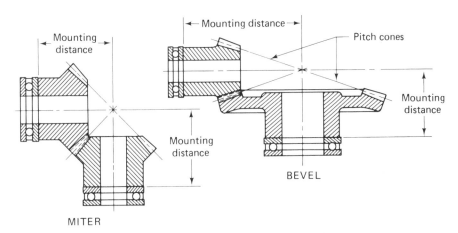

Figure 6.5

In some applications, it is advantageous to have bevel gears in which the axes do not lie in a common plane (offset bevel gears). This permits both shafts to continue beyond their intersection. These gears have teeth of special curvature to make this possible. Examples are *hypoid* and *spiroid* gears. Their action lies between that of bevel gears and of worm gears (see next section).

## 6.3 WORM GEARS

The driving worm has one or more teeth in the shape of a helix, which may be single, double, triple, or quadruple, in similarity to screw threads. Most worms used in industry are either single or double. The speed ratio of a single worm gear drive is $N_g/1$, in which $N_g$ is the number of teeth in the gear. The speed ratio of a double worm and gear is $N_g/2$, and so on for triple and quadruple worms.

Normal pressure angles (i.e., the pressure angle perpendicular to the flank of the helix) may be $14.5°$, $20°$, or $25°$.

If the gear periphery matches that of the worm, the drive is called *semi-enveloping* (see Fig. 6.6). If, in addition, the worm outline matches that of the gear, the drive is said to be *double-enveloping* (see Fig. 6.8). The latter construction provides greater contact area and consequently greater power transmission capability. Since in worm gearing relative velocities and tooth pressures are high, special attention should be given to surface finish and lubrication in order to achieve high efficiency. Worms develop thrust and their bearings must be designed accordingly.

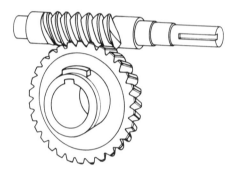

**Figure 6.6**

## 6.4  MATERIALS AND METHODS USED IN MANUFACTURING GEARS

Larger gears may be made from cast iron, steel, or bronze either by the hobbing, shaping, milling, grinding, and casting processes or by any combination thereof. Gear blanks may also be forged, and very large gears may be assembled by welding constituent parts together.

Smaller gears may be made, in addition to the previously mentioned methods, by sintering, plastic molding, stamping from sheet metal, and extrusion.

## 6.5  GEAR TRAINS

When the output shaft of two mating gears, or the driven gear itself, carries a pinion mating with another gear, and possibly when other pairs of gears are added in a similar manner, the entire series of mating gears constitutes a *gear train*.

Two basic gear trains are the *regular* and the *reverted* types [schematically shown in Fig. 6.7 (a) and (b)]. In a reverted gear train the sums of the pitch radii of sets of mating gears must be constant since the center distance is constant (all gears rotate around one of two fixed shafts). The regular gear train is the one most encountered in industry. For both regular and reverted gear trains, the basic overall speed ratio is the product of the tooth numbers of all driv*en* gears divided by the product of the tooth numbers of all driv*er* gears, or

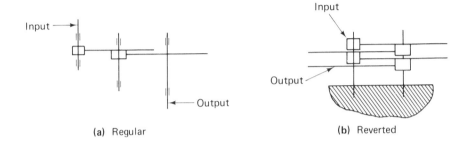

(a) Regular                                                                (b) Reverted

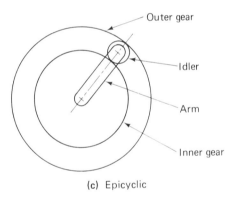

(c) Epicyclic

**Figure 6.7**

$$\text{Speed ratio } (M\omega) = \frac{N_2 \times N_4 \times N_6 \times N_8}{N_1 \times N_3 \times N_5 \times N_7} \text{ etc.}$$

in which $N_{\text{even}}$ are the tooth numbers of the driven gears and $N_{\text{odd}}$ are the tooth numbers of the driving gears.

*Epicyclic gear trains* are compact and use few gears to provide large speed ratios. The basic one is shown schematically in Fig. 6.7(c). One of the three elements (outer gear, arm, or inner gear) is held stationary, and either of the two remaining elements acts either as driver or as follower. Epicyclic gears trains are slightly more expensive to manufacture than the regular and reverted gear trains and for that reason are relatively little used

**Figure 6.8**
Winsmith Div. of UMC.

as industrial gear trains. Fig. 6.8 shows a cutaway view of a special type of epicyclic gear train with two outer gears, one fixed and one connected to the driven (heavy) shaft. The driver shaft carries three parallel sets of two pinions each in a star-shaped configuration. The pinions on the near side engage the fixed outer gear, while the others engage the output gear. The computation of the speed ratio of epicyclic gear trains is beyond the scope of this text.

## 6.6  SPEED REDUCERS

The installation of shafts carrying mating gears has exacting center distance and parallelness requirements. Provisions must also be made for proper lubrication of teeth and bearings and for the removal of frictional heat. For these reasons it is, in most cases, more convenient and economical for prospective users to purchase ready-made gear trains, complete with housing and lubricating system, from a manufacturer, than to assemble gear trains themselves.

Since most gear trains are used to reduce instead of increase speed, such ready-made units are often called *speed reducers*. Every speed reducer has a sump containing lubricating oil into which usually at least one of each pair of mating gears is partially submerged. During operation, this arrangement provides splash lubrication for teeth and bearings. The oil also acts as a transmission device for frictional heat. In larger units an oil pump forces oil through a perforated tube directly on the contact area of the gear teeth. The heat is transmitted by the oil to the housing, which loses it to the surroundings by *radiation* and *convection*. The housing often has cooling ribs to increase the surface area available for convection. When the heat loss by natural convection and radiation is not adequate to remove all the heat at the rated temperature increase of 70°F to 90°F above ambient, as may be the case when the unit operates near the upper limit of its capability, or due to inadequate clearance around the housing, it may be necessary to add a shaft-mounted fan to provide forced airflow across housing walls and ribs. It is also possible to install a *cooling coil* in the sump through which cooling water may be circulated. In large units the oil is often pumped through a *heat exchanger*.

*Backstops* prevent reverse rotation of the unit, which may be necessary or desirable in applications such as hoists or conveyor belt drives. Backstops are available as optional extras from most manufacturers.

Reducers are often available as a unit including an electric or hydraulic motor drive, in which case they are called motorized speed reducers or *gear motors*.

The *efficiency* of speed reducers varies with quality. The best ones may have a power loss of approximately 1 to 1.5% per step. The larger units tend toward the lower value. The overall efficiency is the product of the individual efficiencies of the steps. If, for instance, in a three-step reducer each step has an efficiency of .99, the unit efficiency is .99 × .99 × .99 = .97.

Although theoretically almost any speed ratio is obtainable in a gear train, the American Gear Manufacturers Association (AGMA) has established standard catalog ratios for commercially available speed reducers. Most manufacturers supply reducers that have special, nonstandard ratios. Of course, these reducers cost more than comparable standard-ratio reducers.

Actual values of standard manufacturers' catalog ratios may vary ±3%.

The following types are in general use: worm gear reducers, parallel shaft reducers, spiral bevel reducers, concentric shaft reducers, and shaft-mounted reducers. They are discussed below.

*WORM GEAR REDUCER.* A worm gear reducer (see Fig. 6.9) provides considerable speed reduction in a single step, and it is often the most compact unit available for a given capacity. Worm gear reducers are commonly available with single or double worms.

A double reduction unit consists of two worm gear reducers in series and provides an overall speed ratio equal to the product of the ratios

**Figure 6.9**
The Falk Corporation Milwaukee, Wisconsin.

of each individual unit. For example, a 20/1 unit connected to a 8/1 unit gives an overall reduction of 160/1.

Worm gear reducers are quiet and vibration-free, while the power output has no pulsations. They are well-suited to service conditions with heavy shock loads due to the inherent strength of the worm teeth. Some manufacturers provide optional torque control to protect both driving and driven equipment. When a predetermined torque value is exceeded, a microswitch is actuated which interrupts the motor control circuit.

Worm gear reducers are sometimes self-locking and may not need backstop devices. Their efficiency is somewhat lower than that of the pinion and gear types.

*PARALLEL SHAFT REDUCER.* As the name implies, this reducer has parallel input and output shafts. Parallel shaft reducers may have one, two, three, or four pinion and gear sets. The pinion of the second set may

74

be mounted on the shaft of the first gear, or it may form one part with the latter, and so on. They are identified as single, double, triple, and quadruple units (see Figs. 6.10 and 6.11 for a triple reduction unit).

The gears are sometimes of the spur type, but in most cases helical gears are used.

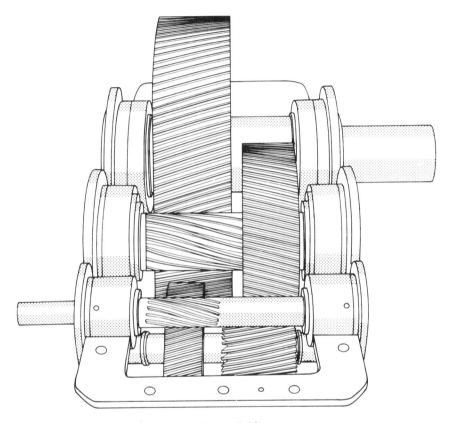

Figure 6.10

*SPIRAL BEVEL REDUCER.* This reducer is very similar to the parallel shaft reducer, except that the first set of reduction gears is of the bevel type (usually spiral bevel gears). In this arrangement input and output shafts are at right angles to one another.

*CONCENTRIC SHAFT REDUCER.* This reducer (see Fig. 6.12) is used where space conditions require that input and output shafts have a common centerline. It is an example of a reverted gear train and is especially well-suited to easy change of speed ratio through the use of standardized components, which are available from manufacturers.

**Figure 6.11**
The Falk Corporation Milwaukee, Wisconsin.

*SHAFT-MOUNTED REDUCER.* This reducer is also called *torque arm reducer* and may save considerable floor space. It has a hollow output spindle that is keyed to the driven equipment's input shaft in lieu of a coupling, sheave, gear, or other power transmission element. The torque arm needed to prevent it from rotation may be designed to actuate a torque-sensitive switch that interrupts the motor control circuit if there is an overload.

Figure 6.13 shows a typical shaft-mounted unit with its cover removed and with some options in torque-arm locations. Torque arms must be under tensile, not compressive, load. Shaft-mounted reducers are typically driven by V-belts, although many have bolted-on electric or hydraulic motors.

## 6.7 SELECTING REDUCERS

After speed ratio and desired shaft configuration have been decided upon, a suitable type of reducer may be selected. The horsepower to be transmitted is multipled by the appropriate service factor and a unit of ade-

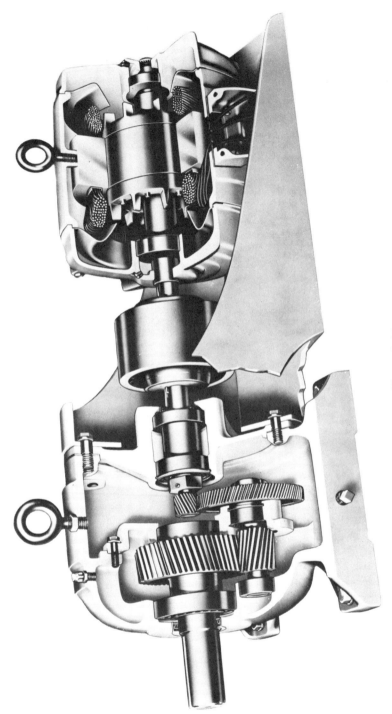

Figure 6.12

PTD, Dresser Industries.

Figure 6.13

Reliance Electric Co., Mishawaka, Indiana

quate rating is selected from a catalog. The application of the service factor ensures the unit's ability to carry temporary overloads.

As mentioned before, the heat generated by friction in a reducer is mostly dissipated to the surroundings by natural radiation and some convection. A given housing size has a certain heat dissipation capability that is proportional to its total exposed surface area, including any cooling fins. Each housing size of a series of units with increasing ratings is used for a certain range of horsepower, for instance, from 160 hp to 200 hp.

The friction loss is a percentage of the power transmitted and is proportional to it. It is clear that near the top of the range (say 195 hp for the housing discussed in the above paragraph) and under heavy-duty conditions, it is possible that the housing will not have enough capacity for the natural dissipation of all the heat generated. (Permissible temperature rise is from 70°F to 90°F above ambient).

It is therefore necessary to compare the *average horsepower transmitted* (i.e., the value *before* the service factor is applied) to the *thermal horsepower* rating of the unit, which is usually provided in the table next to the one giving mechanical ratings. The thermal horsepower is the maximum amount of power that can be transmitted on a *continuous* basis with the temperature rise not exceeding 70°F to 90°F above ambient. (Maximum oil temperature permitted is usually 160°F.) If the thermal horsepower value found in the table is below the horsepower to be transmitted, forced cooling means have to be provided, such as cooling fans, cooling coils, or an oil cooler with a circulation pump. An alternate solution is to investigate whether the next larger unit has an adequate thermal horsepower capacity for the load.

Another item to be checked is the *overhung load*. This is the reaction force caused by sprockets, pinions or gears, sheaves, or pulleys, which are mounted either on the input or on the output shaft. The equivalent overhung load $P$ (pounds) is found from

$$P = \frac{\text{hp} \times 63{,}000 \times K}{\text{rpm} \times R_p}$$

in which hp is horsepower *transmitted* (before the service factor is applied), $K$ is a factor depending on the width of the sprocket, pinion, sheave, or pulley used, rpm is revolutions per minute of the shaft, and $R_p$ is the pitch *radius* of the sprocket, pinion, sheave, or pulley. $K$-factors are given in the following table:

TABLE 6.1

| Overhung element | K-factor |
|---|---|
| Sprocket | 1 |
| Spur pinion | 1.25 |
| V-belt sheave | 1.5 |
| Flat belt pulley | 2.5 |

The calculated load $P$ should not exceed the tabulated values in the catalog table of maximum permissible overhung loads. If the overhung load calculated is greater than that permitted, the following remedial measures are available:

1.  Increase the pitch radius of the drive element.
2.  Use a larger reducer.
3.  Provide an outboard bearing.

Some catalogs also permit selection of reducers on the basis of required output torque.

The following example will clarify the selection procedure on the basis of horsepower.

> **Given:** A parallel shaft reducer is to transmit 95 hp maximum. The driver is a direct-connected, 1,750 rpm electric motor, and the driven equipment is a multicylinder, reciprocating compressor running at 375 rpm 24 h/day.
>
> **Required:** To find a suitable speed reducer.
>
> **Solution:** It has been established that the service factor for this application is 1.5. The rated horsepower required is then 1.5 × 95 = 142.5. The speed ratio is 1750/375 = 4.66.

Go to the table reproduced in Fig. 6.14. The speed ratio indicates a single-reduction unit, type HLE. The nearest standard ratio is found to be 5.062. If this is acceptable, the smallest available unit proves to be type 0901. This unit has a mechanical (rated) hp capacity of 235, which is more than adequate. The thermal horsepower is 141, which is more than 95 and is thus acceptable.

The unit will have a direct-connected driver motor, but it will drive the compressor via V-belts. Therefore, we must check the overhung load on the output side.

If the output sheave has a pitch diameter of 10 in. and if the point of application of the reaction load (i.e., the center of the combined V-belts) lies over the center of the keyway, you can now apply the equation shown on page 79.

$$P = \frac{95 \times 63,000 \times 1.5}{346 \times 5} = 5,200 \text{ lb}$$

The table of overhung loads in Fig. 6.14 indicates a permissible value of 9,550 lb. The unit is therefore suitable.

### NOTES

1. If the point of application of the overhung load lies outside the center of the keyway, the overhung load must be obtained from the factory's representative.

2. The revolutions per minute value of 346 rather than 375 reflects the different speed ratio used.

## 1750 RPM INPUT SPEED RATINGS

| REDUCER SIZE | RATIO | | NOMINAL OUTPUT RPM | INPUT HP | | OUTPUT TORQUE inch/lbs | OVERHUNG LOAD (LBS.) | | |
|---|---|---|---|---|---|---|---|---|---|
| | NOMINAL | ACTUAL | | MECH. | THERMAL 1 & 2 | | OUTPUT SHAFT | | INPUT SHAFT |
| | | | | | | | STYLE 1[3] | STYLE 4[3] | |
| 0901 | 1.837 | 1.837 | 950 | 800 | 113 | 52100 | 5870 | — | 1570 |
| | 2.250 | 2.237 | 780 | 675 | 120 | 53500 | 6520 | — | 2370 |
| | 2.756 | 2.784 | 635 | 527 | 120 | 52000 | 7180 | — | 2560 |
| | 3.375 | 3.375 | 520 | 415 | 125 | 49600 | 7920 | — | 2840 |
| | 4.134 | 4.125 | 425 | 326 | 132 | 47700 | 8730 | — | * |
| | 5.062 | 5.087 | 346 | 235 | 141 | 42400 | 9550 | — | 960 |
| | 6.200 | 6.182 | 282 | 175 | 150 | 38400 | 10400 | — | 1330 |
| | 7.594 | 7.476 | 230 | 137 | — | 36300 | 11200 | — | 1700 |
| | 1.837 | 1.829 | 950 | 1050 | 125 | 67800 | 6170 | — | 2870 |
| | 2.250 | 2.238 | 780 | 900 | 132 | 71500 | 6860 | — | 3070 |

Figure 6.14

PTD, Dresser Industries.

## PROBLEMS

1. Make a quarter-scale layout of the drive (type HLE, size 0901) selected on pages 79 and 80 of this chapter. The sketches in Figs. 6.15 and 6.16 and dimension tables (Tables 6.2 and 6.3) are typical. They are generally not suited for construction purposes, for which certified prints should be used. Note that reducer dimensions are given in Table 6.2 and bedplate dimensions are given in Table 6.3. Use size C vellum and show a front view and a top view on two separate sheets. Scale the dimensions of the electric motor (a standard NEMA frame size) and any other dimensions not tabulated. Draw the bedplate to finished, not rough, bottom dimensions (see Note 1 in Table 6.3 for dimensions C, D, E, and F). On your drawing call out all dimensions found in these tables.

2. What are the advantages and disadvantages of helical gearing?

3. What is the purpose of offset bevel gearing?

4. Why is the surface finish of particular importance in worm gearing?

5. Explain the difference between rated (mechanical) horsepower and thermal horsepower.

6. Mention three ways to increase the thermal horsepower rating of a reducer.

7. What is meant by an overhung load? Given a certain amount of horsepower to be transmitted, how can the overhung load be reduced?

8. What is the purpose of a backstop? Does every reducer need a backstop?

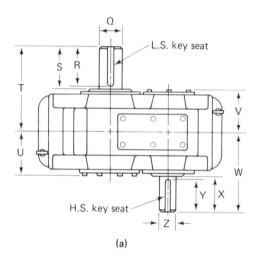

(a)

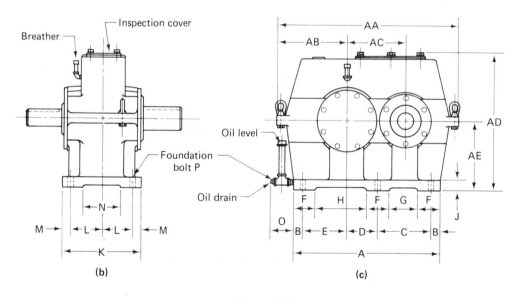

(b)

(c)

**Figure 6.15**
PTD, Dresser Industries.

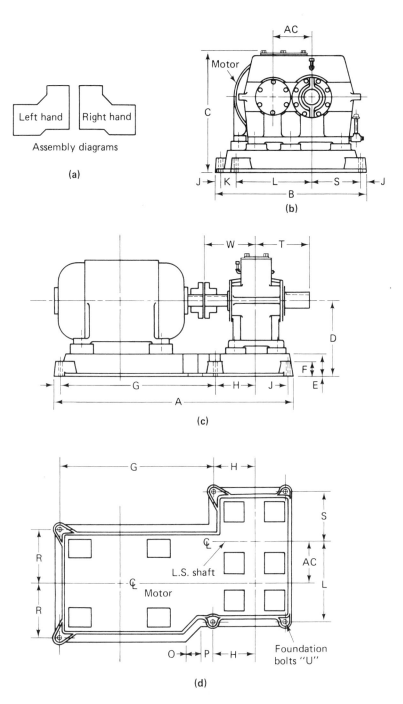

Figure 6.16
PTD, Dresser Industries.

TABLE 6.2

# PARALLEL SHAFT DIMENSIONS
## TYPE HLE SINGLE REDUCTION

**85**

| REDUCER SIZE | AA | AB | AC | AD | AE | A | B | C | D | E | F | G |
|---|---|---|---|---|---|---|---|---|---|---|---|---|
| 0701 | 23.12 | 9.06 | 7 | 16.38 | 8.25 | 18.75 | 2.38 | 5.75 | 3.75 | 4.50 | 4.00 | 2.12 |
| 0801 | 25.88 | 10.06 | 8 | 18.62 | 9.50 | 20.75 | 2.50 | 6.75 | 3.88 | 5.12 | 4.25 | 2.75 |
| 0901 | 28.38 | 11.06 | 9 | 20.88 | 10.50 | 23.25 | 2.50 | 7.50 | 4.75 | 6.00 | 4.50 | 3.25 |
| 1001 | 31.38 | 12.31 | 10 | 22.88 | 11.50 | 26.50 | 2.75 | 8.50 | 5.38 | 7.12 | 5.00 | 3.75 |
| 1101 1201 | 36.88 | 14.56 | 12 | 27.25 | 13.75 | 30.25 | 3.00 | 10.00 | 6.00 | 8.25 | 5.50 | 4.75 |
| 1301 1401 | 42.62 | 17.06 | 14 | 31.25 | 15.75 | 36.00 | 3.25 | 11.50 | 7.50 | 10.50 | 6.00 | 5.75 |
| 1501 1601 | 47.75 | 19.12 | 16 | 35.62 | 18.00 | 41.50 | 3.75 | 13.62 | 8.25 | 12.12 | 7.00 | 6.88 |
| 1701 1801 | 55.00 | 21.88 | 18 | 39.62 | 20.00 | 46.25 | 4.25 | 15.00 | 9.25 | 13.50 | 8.00 | 7.25 |
| 1901 2001 | 59.50 | 23.50 | 20 | 44.38 | 23.00 | 49.00 | 5.25 | 15.75 | 9.75 | 13.00 | 8.25 | 8.62 |

| REDUCER SIZE | H | J | K | L | M | N | O | P | LOW SPEED SHAFT | | R | S |
|---|---|---|---|---|---|---|---|---|---|---|---|---|
| | | | | | | | | | Q⊙ | KEY SEAT | | |
| 0701 | 4.62 | 1.38 | 11.50 | 4.75 | 1.00 | 6.00 | 4.25 | .88 | 2.88 | .75 x .38 | 5.00 | 5.25 |
| 0801 | 5.25 | 1.50 | 12.50 | 5.12 | 1.12 | 6.50 | 4.50 | 1.00 | 3.38 | 1.00 x .50 | 5.94 | 6.19 |
| 0901 | 6.50 | 1.62 | 13.50 | 5.62 | 1.12 | 7.00 | 4.62 | 1.00 | 3.88 | 1.00 x .50 | 6.62 | 6.88 |
| 1001 | 7.75 | 1.75 | 14.00 | 5.75 | 1.25 | 7.00 | 4.62 | 1.12 | 4.50 | 1.25 x .62 | 7.38 | 7.62 |
| 1101 1201 | 9.00 | 2.00 | 15.50 | 6.38 | 1.38 | 8.50 | 5.50 | 1.25 | 5.00 | 1.25 x .62 | 8.50 | 8.75 |
| 1301 1401 | 12.25 | 2.25 | 17.50 | 7.38 | 1.38 | 9.50 | 5.50 | 1.25 | 6.00 | 1.50 x .75 | 9.75 | 10.00 |

TABLE 6.2 (*Continued*)

| REDUCER SIZE | H | J | K | L | M | N | O | P | LOW SPEED SHAFT Q⊙ | KEY SEAT | R | S |
|---|---|---|---|---|---|---|---|---|---|---|---|---|
| 1501 1601 | 13.62 | 2.50 | 19.25 | 8.00 | 1.62 | 10.25 | 5.75 | 1.50 | 6.50 | 1.50 x .75 | 10.50 | 10.75 |
| 1701 1801 | 15.00 | 2.75 | 22.00 | 9.12 | 1.88 | 12.50 | 6.75 | 1.75 | 7.25 | 1.75 x .62 | 11.38 | 11.62 |
| 1901 2001 | 15.62 | 3.00 | 23.50 | 9.62 | 2.12 | 13.50 | 7.88 | 2.00 | 7.75 | 2.00 x .75 | 11.75 | 12.12 |

| REDUCER SIZE | T | U | V | W | X | Y | H.S. SHAFTS FOR RATIOS 1.837 - 3.375 Z⊙ | KEY SEAT | 4.134 - 7.594 Z⊙ | KEY SEAT | APPROX. WT/LBS |
|---|---|---|---|---|---|---|---|---|---|---|---|
| 0701 | 11.75 | 6.50 | 6.50 | 11.75 | 5.25 | 5.00 | 2.12 | .50 x .25 | 1.75 | .38 x .19 | 500 |
| 0801 | 13.25 | 7.25 | 7.12 | 13.25 | 6.19 | 5.94 | 2.38 | .62 x .31 | 1.88 | .50 x .25 | 650 |
| 0901 | 14.50 | 7.75 | 7.62 | 14.25 | 6.69 | 6.44 | 2.62 | .62 x .31 | 2.00 | .50 x .25 | 780 |
| 1001 | 15.50 | 8.00 | 8.00 | 15.50 | 7.62 | 7.38 | 3.00 | .75 x .38 | 2.25 | .50 x .25 | 1175 |
| 1101 1201 | 17.50 | 8.75 | 8.88 | 17.25 | 8.50 | 8.25 | 3.38 | .88 x .44 | 2.50 | .62 x .31 | 1540 |
| 1301 1401 | 19.75 | 10.00 | 10.00 | 19.00 | 9.25 | 9.00 | 3.75 | 1.00 x .50 | 3.00 | .75 x .38 | 1975 |
| 1501 1601 | 21.50 | 10.88 | 10.88 | 21.50 | 10.75 | 10.50 | 4.50 | 1.25 x .62 | 3.25 | .75 x .38 | 3250 |
| 1701 1801 | 24.25 | 12.50 | 12.38 | 24.00 | 11.38 | 11.00 | 5.00 | 1.25 x .62 | 4.00 | 1.00 x .50 | 4450 |
| 1901 2001 | 25.50 | 13.50 | 13.25 | 25.00 | 11.62 | 11.25 | 5.62 | 1.50 x .75 | 4.00 | 1.00 x .50 | 5950 |

⊙ Shaft Diameters of 2 inches and under are held to limits of +.0000" / −.0005"

Shaft Diameters over 2 inches are held to limits of +.000" / −.001"

(Courtesy: PTD, Dresser Industries.)

TABLE 6.3

# 100  PARALLEL SHAFT BEDPLATES
## TYPE HLE SINGLE REDUCTION

| Reducer Size | Motor Frame | A | B | C | D | E | F | G | H* | J | K | L | M | N | P |
|---|---|---|---|---|---|---|---|---|---|---|---|---|---|---|---|
| 1001 | 404 | 27.80 | 3.38 | 8.50 | 12.50 | 3.88 | 2.00 | 5.75 | 1.00 | 6-1.12 | 18.00 | 2.40 | 1.00 | 11.50 | 25.00 |
|  | 405 | 27.80 | 3.38 | 8.50 | 12.50 | 3.88 | 2.00 | 5.75 | 1.00 | 6-1.12 | 18.00 | 2.40 | 1.00 | 11.50 | 25.00 |
|  | 444 | 27.80 | 3.38 | 8.50 | 12.50 | 3.88 | 2.00 | 5.75 | 1.00 | 6-1.12 | 20.70 | 2.40 | 1.00 | 13.00 | 28.00 |
|  | 445 | 27.80 | 3.38 | 8.50 | 12.50 | 3.88 | 2.00 | 5.75 | 1.00 | 6-1.12 | 20.70 | 2.40 | 1.00 | 13.00 | 28.00 |
|  | 504 | 27.80 | 3.38 | 8.50 | 12.50 | 3.88 | 2.00 | 5.75 | 1.00 | 6-1.12 | 23.75 | 1.12 | 2.50 | 10.00 | 25.00 |
|  | 505 | 27.80 | 3.38 | 8.50 | 12.50 | 3.88 | 2.00 | 5.75 | 1.00 | 6-1.12 | 25.75 | 1.12 | 2.50 | 10.00 | 25.00 |
|  | 507 | 27.80 | 3.38 | 8.50 | 12.50 | 3.88 | 2.00 | 5.75 | 1.00 | 6-1.12 | 29.75 | 1.12 | 2.50 | 10.00 | 25.00 |
|  | 509 | 27.80 | 3.38 | 8.50 | 12.50 | 3.88 | 2.00 | 5.75 | 1.00 | 6-1.12 | 35.75 | 1.12 | 2.50 | 10.00 | 25.00 |
|  | 587 | 27.80 | 3.38 | 8.50 | 12.50 | 3.88 | 2.00 | 5.75 | 1.00 | 6-1.12 | 33.00 | 1.12 | 2.50 | 12.25 | 29.50 |
|  | 588 | 27.80 | 3.38 | 8.50 | 12.50 | 3.88 | 2.00 | 5.75 | 1.00 | 6-1.12 | 36.00 | 1.12 | 2.50 | 12.25 | 29.50 |
|  | 589 | 27.80 | 3.38 | 8.50 | 12.50 | 3.88 | 2.00 | 5.75 | 1.00 | 6-1.12 | 40.00 | 1.12 | 2.50 | 12.25 | 29.50 |
|  | 687 | 27.80 | 3.38 | 8.50 | 12.50 | 3.88 | 2.00 | 5.75 | 1.00 | 6-1.12 | 41.75 | 1.12 | 2.50 | 14.25 | 33.50 |
|  | 688 | 27.80 | 3.38 | 8.50 | 12.50 | 3.88 | 2.00 | 5.75 | 1.00 | 6-1.12 | 45.75 | 1.12 | 2.50 | 14.25 | 33.50 |
| 1101 1201 | 504US | 31.50 | 3.62 | 10.00 | 14.25 | 4.00 | 2.38 | 6.38 | 1.00 | 6-1.12 | 23.75 | 1.12 | 2.50 | 10.00 | 25.00 |
|  | 505US | 31.50 | 3.62 | 10.00 | 14.25 | 4.00 | 2.38 | 6.38 | 1.00 | 6-1.12 | 25.75 | 1.12 | 2.50 | 10.00 | 25.00 |
|  | 507US | 31.50 | 3.62 | 10.00 | 14.25 | 4.00 | 2.38 | 6.38 | 1.00 | 6-1.12 | 29.75 | 1.12 | 2.50 | 10.00 | 25.00 |
|  | 509US | 31.50 | 3.62 | 10.00 | 14.25 | 4.00 | 2.38 | 6.38 | 1.00 | 6-1.12 | 35.75 | 1.12 | 2.50 | 10.00 | 25.00 |
|  | 587S | 31.50 | 3.62 | 10.00 | 14.25 | 4.00 | 2.38 | 6.38 | 1.00 | 6-1.12 | 33.00 | 1.12 | 2.50 | 12.25 | 29.50 |
|  | 588S | 31.50 | 3.62 | 10.00 | 14.25 | 4.00 | 2.38 | 6.38 | 1.00 | 6-1.12 | 36.00 | 1.12 | 2.50 | 12.25 | 29.50 |
|  | 589S | 31.50 | 3.62 | 10.00 | 14.25 | 4.00 | 2.38 | 6.38 | 1.00 | 6-1.12 | 40.00 | 1.12 | 2.50 | 12.25 | 29.50 |
|  | 687S | 31.50 | 3.62 | 10.00 | 14.25 | 4.00 | 2.38 | 6.38 | 1.00 | 6-1.12 | 41.75 | 1.12 | 2.50 | 14.25 | 33.50 |
|  | 688S | 31.50 | 3.62 | 10.00 | 14.25 | 4.00 | 2.38 | 6.38 | 1.00 | 6-1.12 | 45.75 | 1.12 | 2.50 | 14.25 | 33.50 |
| 1301 1401 | 507US | 37.30 | 3.88 | 11.50 | 18.00 | 5.00 | 2.38 | 7.38 | 1.25 | 6-1.25 | 29.75 | 1.12 | 2.50 | 10.00 | 25.00 |
|  | 509US | 37.30 | 3.88 | 11.50 | 18.00 | 5.00 | 2.38 | 7.38 | 1.25 | 6-1.25 | 35.75 | 1.12 | 2.50 | 10.00 | 25.00 |
|  | 587S | 37.30 | 3.88 | 11.50 | 18.00 | 5.00 | 2.38 | 7.38 | 1.25 | 6-1.25 | 33.00 | 1.12 | 2.50 | 12.25 | 29.50 |
|  | 588S | 37.30 | 3.88 | 11.50 | 18.00 | 5.00 | 2.38 | 7.38 | 1.25 | 6-1.25 | 36.00 | 1.12 | 2.50 | 12.25 | 29.50 |
|  | 589S | 37.30 | 3.88 | 11.50 | 18.00 | 5.00 | 2.38 | 7.38 | 1.25 | 6-1.25 | 40.00 | 1.12 | 2.50 | 12.25 | 29.50 |
|  | 687S | 37.30 | 3.88 | 11.50 | 18.00 | 5.00 | 2.38 | 7.38 | 1.25 | 6-1.25 | 41.75 | 1.12 | 2.50 | 14.25 | 33.50 |
|  | 688S | 37.30 | 3.88 | 11.50 | 18.00 | 5.00 | 2.38 | 7.38 | 1.25 | 6-1.25 | 45.75 | 1.12 | 2.50 | 14.25 | 33.50 |
| 1501 1601 | 509US | 42.80 | 4.38 | 13.62 | 20.38 | 5.88 | 2.50 | 8.00 | 1.25 | 6-1.50 | 35.75 | 1.12 | 2.50 | 10.00 | 25.00 |
|  | 587S | 42.80 | 4.38 | 13.62 | 20.38 | 5.88 | 2.50 | 8.00 | 1.25 | 6-1.50 | 33.00 | 1.12 | 2.50 | 12.25 | 29.50 |
|  | 588S | 42.80 | 4.38 | 13.62 | 20.38 | 5.88 | 2.50 | 8.00 | 1.25 | 6-1.50 | 36.00 | 1.12 | 2.50 | 12.25 | 29.50 |
|  | 589S | 42.80 | 4.38 | 13.62 | 20.38 | 5.88 | 2.50 | 8.00 | 1.25 | 6-1.50 | 40.00 | 1.12 | 2.50 | 12.25 | 29.50 |
|  | 687S | 42.80 | 4.38 | 13.62 | 20.38 | 5.88 | 2.50 | 8.00 | 1.25 | 6-1.50 | 41.75 | 1.12 | 2.50 | 14.25 | 33.50 |
|  | 688S | 42.80 | 4.38 | 13.62 | 20.38 | 5.88 | 2.50 | 8.00 | 1.25 | 6-1.50 | 45.75 | 1.12 | 2.50 | 14.25 | 33.50 |
|  | 689S | 42.80 | 4.38 | 13.62 | 20.38 | 5.88 | 2.50 | 8.00 | 1.25 | 6-1.50 | 47.75 | 1.12 | 2.50 | 14.25 | 33.50 |
| 1701 1801 | 588S | 47.50 | 4.88 | 15.00 | 22.75 | 6.25 | 2.75 | 9.12 | 1.25 | 6-1.75 | 40.00 | 1.12 | 2.50 | 12.25 | 29.50 |
|  | 687S | 47.50 | 4.88 | 15.00 | 22.75 | 6.25 | 2.75 | 9.12 | 1.25 | 6-1.75 | 41.75 | 1.12 | 2.50 | 14.25 | 33.50 |
|  | 688S | 47.50 | 4.88 | 15.00 | 22.75 | 6.25 | 2.75 | 9.12 | 1.25 | 6-1.75 | 45.75 | 1.12 | 2.50 | 14.25 | 33.50 |
|  | 689S | 47.50 | 4.88 | 15.00 | 22.75 | 6.25 | 2.75 | 9.12 | 1.25 | 6-1.75 | 47.75 | 1.12 | 2.50 | 14.25 | 33.50 |
| 1901 2001 | 687S | 50.30 | 5.88 | 15.75 | 22.75 | 5.50 | 2.88 | 9.62 | 1.25 | 6-2.00 | 41.75 | 1.12 | 2.50 | 14.25 | 33.50 |
|  | 688S | 50.30 | 5.88 | 15.75 | 22.75 | 5.50 | 2.88 | 9.62 | 1.25 | 6-2.00 | 45.75 | 1.12 | 2.50 | 14.25 | 33.50 |
|  | 689S | 50.30 | 5.88 | 15.75 | 22.75 | 5.50 | 2.88 | 9.62 | 1.25 | 6-2.00 | 47.75 | 1.12 | 2.50 | 14.25 | 33.50 |

TABLE 6.3 *(Continued)*

| Reducer Size | Motor Frame | Q | R | S* | T | U | W | X | AC | BA | BB | BC* | BD* | BE | BF | Approx. Weight Lbs. |
|---|---|---|---|---|---|---|---|---|---|---|---|---|---|---|---|---|
| 1001 | 404 | 22.80 | 4-1.00 | 1.25 | 15.50 | 3.90 | 15.50 | 4.50 | 10.00 | 33.00 | 53.80 | 23.00 | 34.40 | 19.90 | 7.62 | 975 |
| | 405 | 22.80 | 4-1.00 | 1.25 | 15.50 | 3.90 | 15.50 | 4.50 | 10.00 | 33.00 | 54.50 | 23.00 | 34.40 | 20.60 | 7.62 | 1015 |
| | 444 | 25.50 | 4-1.00 | 1.25 | 15.50 | 3.90 | 15.50 | 4.50 | 10.00 | 34.50 | 58.00 | 23.00 | 34.40 | 21.40 | 9.12 | 1050 |
| | 445 | 25.50 | 4-1.00 | 1.25 | 15.50 | 3.90 | 15.50 | 4.50 | 10.00 | 34.50 | 59.00 | 23.00 | 34.40 | 22.40 | 9.12 | 1095 |
| | 504 | 26.00 | 4-1.00 | 1.25 | 15.50 | 2.75 | 15.50 | 4.50 | 10.00 | 33.00 | 58.10 | 23.00 | 34.40 | 19.75 | 6.12 | 1135 |
| | 505 | 28.00 | 4-1.00 | 1.25 | 15.50 | 2.75 | 15.50 | 4.50 | 10.00 | 33.00 | 60.10 | 23.00 | 34.40 | 19.75 | 6.12 | 1170 |
| | 507 | 32.00 | 4-1.00 | 1.25 | 15.50 | 2.75 | 15.50 | 4.50 | 10.00 | 33.00 | 64.10 | 23.00 | 34.40 | 19.75 | 6.12 | 1225 |
| | 509 | 38.00 | 4-1.00 | 1.25 | 15.50 | 3.00 | 15.50 | 4.50 | 10.00 | 33.00 | 70.10 | 23.00 | 34.40 | 19.75 | 6.12 | 1265 |
| | 587 | 35.80 | 4-1.00 | 1.25 | 15.50 | 3.00 | 15.50 | 4.50 | 10.00 | 35.30 | 68.80 | 23.00 | 34.40 | 21.12 | 8.38 | 1290 |
| | 588 | 38.80 | 4-1.00 | 1.25 | 15.50 | 3.00 | 15.50 | 4.50 | 10.00 | 35.30 | 71.80 | 23.00 | 34.40 | 21.12 | 8.38 | 1335 |
| | 589 | 42.80 | 4-1.00 | 1.25 | 15.50 | 3.00 | 15.50 | 4.50 | 10.00 | 35.30 | 74.80 | 23.00 | 34.40 | 20.12 | 8.38 | 1380 |
| | 687 | 44.00 | 4-1.00 | 1.25 | 15.50 | 3.50 | 15.50 | 4.50 | 10.00 | 37.30 | 79.10 | 23.00 | 34.40 | 22.75 | 10.38 | 1415 |
| | 688 | 48.00 | 4-1.00 | 1.25 | 15.50 | 3.50 | 15.50 | 4.50 | 10.00 | 37.30 | 83.10 | 23.00 | 34.40 | 22.75 | 10.38 | 1475 |
| 1101 1201 | 504US | 26.00 | 4-1.00 | 1.25 | 17.50 | 2.75 | 17.25 | 4.80 | 12.00 | 36.40 | 60.90 | 23.00 | 36.50 | 20.87 | 6.00 | 1075 |
| | 505US | 28.00 | 4-1.00 | 1.25 | 17.50 | 2.75 | 17.25 | 4.80 | 12.00 | 36.40 | 62.90 | 23.00 | 36.50 | 20.87 | 6.00 | 1125 |
| | 507US | 32.00 | 4-1.00 | 1.25 | 17.50 | 2.75 | 17.25 | 4.80 | 12.00 | 36.40 | 66.90 | 23.00 | 36.50 | 20.87 | 6.00 | 1175 |
| | 509US | 38.00 | 4-1.00 | 1.25 | 17.50 | 3.00 | 17.25 | 4.80 | 12.00 | 36.40 | 72.90 | 23.00 | 36.50 | 20.87 | 6.00 | 1225 |
| | 587S | 35.80 | 4-1.00 | 1.25 | 17.50 | 3.00 | 17.25 | 4.80 | 12.00 | 38.60 | 71.50 | 23.00 | 36.50 | 22.25 | 8.25 | 1275 |
| | 588S | 38.80 | 4-1.00 | 1.25 | 17.50 | 3.00 | 17.25 | 4.80 | 12.00 | 38.60 | 74.50 | 23.00 | 36.50 | 22.25 | 8.25 | 1325 |
| | 589S | 42.80 | 4-1.00 | 1.25 | 17.50 | 3.00 | 17.25 | 4.80 | 12.00 | 38.60 | 77.50 | 23.00 | 36.50 | 21.25 | 8.25 | 1375 |
| | 687S | 44.00 | 4-1.00 | 1.25 | 17.50 | 3.50 | 17.25 | 4.80 | 12.00 | 40.60 | 81.90 | 23.00 | 36.50 | 23.88 | 10.25 | 1425 |
| | 688S | 48.00 | 4-1.00 | 1.25 | 17.50 | 3.50 | 17.25 | 4.80 | 12.00 | 40.60 | 85.90 | 23.00 | 36.50 | 23.88 | 10.25 | 1475 |
| 1301 1401 | 507US | 32.00 | 4-1.00 | 1.25 | 19.75 | 2.75 | 19.00 | 5.30 | 14.00 | 40.90 | 69.60 | 25.00 | 40.50 | 21.62 | 5.00 | 1475 |
| | 509US | 38.00 | 4-1.00 | 1.25 | 19.75 | 3.00 | 19.00 | 5.30 | 14.00 | 40.90 | 75.60 | 25.00 | 40.50 | 21.62 | 5.00 | 1540 |
| | 587S | 35.80 | 4-1.00 | 1.25 | 19.75 | 3.00 | 19.00 | 5.30 | 14.00 | 43.10 | 74.30 | 25.00 | 40.50 | 23.00 | 7.25 | 1605 |
| | 588S | 38.80 | 4-1.00 | 1.25 | 19.75 | 3.00 | 19.00 | 5.30 | 14.00 | 43.10 | 77.30 | 25.00 | 40.50 | 23.00 | 7.25 | 1700 |
| | 589S | 42.80 | 4-1.00 | 1.25 | 19.75 | 3.00 | 19.00 | 5.30 | 14.00 | 43.10 | 81.30 | 25.00 | 40.50 | 23.00 | 7.25 | 1760 |
| | 687S | 44.00 | 4-1.00 | 1.25 | 19.75 | 3.50 | 19.00 | 5.30 | 14.00 | 45.10 | 84.60 | 25.00 | 40.50 | 24.62 | 9.25 | 1810 |
| | 688S | 48.00 | 4-1.00 | 1.25 | 19.75 | 3.50 | 19.00 | 5.30 | 14.00 | 45.10 | 88.60 | 25.00 | 40.50 | 24.62 | 9.25 | 1925 |
| 1501 1601 | 509US | 38.00 | 4-1.00 | 1.25 | 21.50 | 3.00 | 21.50 | 5.80 | 16.00 | 45.00 | 76.40 | 27.00 | 44.60 | 21.00 | 4.12 | 1925 |
| | 587S | 35.80 | 4-1.00 | 1.25 | 21.50 | 3.00 | 21.50 | 5.80 | 16.00 | 47.30 | 77.50 | 27.00 | 44.60 | 24.87 | 6.38 | 2005 |
| | 588S | 38.80 | 4-1.00 | 1.25 | 21.50 | 3.00 | 21.50 | 5.80 | 16.00 | 47.30 | 80.50 | 27.00 | 44.60 | 24.87 | 6.38 | 2085 |
| | 589S | 42.80 | 4-1.00 | 1.25 | 21.50 | 3.00 | 21.50 | 5.80 | 16.00 | 47.30 | 83.50 | 27.00 | 44.60 | 23.75 | 6.38 | 2175 |
| | 687S | 44.00 | 4-1.00 | 1.25 | 21.50 | 3.50 | 21.50 | 5.80 | 16.00 | 49.30 | 87.90 | 27.00 | 44.60 | 26.50 | 8.38 | 2250 |
| | 688S | 48.00 | 4-1.00 | 1.25 | 21.50 | 3.50 | 21.50 | 5.80 | 16.00 | 49.30 | 91.90 | 27.00 | 44.60 | 26.50 | 8.38 | 2330 |
| | 689S | 50.00 | 4-1.00 | 1.25 | 21.50 | 3.50 | 21.50 | 5.80 | 16.00 | 49.30 | 94.90 | 27.00 | 44.60 | 27.50 | 8.38 | 2425 |
| 1701 1801 | 589S | 42.80 | 4-1.00 | 1.25 | 24.25 | 3.00 | 24.00 | 6.30 | 18.00 | 51.10 | 87.40 | 29.00 | 48.60 | 25.25 | 6.00 | 2425 |
| | 687S | 44.00 | 4-1.00 | 1.25 | 24.25 | 3.50 | 24.00 | 6.30 | 18.00 | 53.10 | 91.80 | 29.00 | 48.60 | 27.88 | 8.00 | 2480 |
| | 688S | 48.00 | 4-1.00 | 1.25 | 24.25 | 3.50 | 24.00 | 6.30 | 18.00 | 53.10 | 95.80 | 29.00 | 48.60 | 27.88 | 8.00 | 2540 |
| | 689S | 50.00 | 4-1.00 | 1.25 | 24.25 | 3.50 | 24.00 | 6.30 | 18.00 | 53.10 | 98.80 | 29.00 | 48.60 | 28.88 | 8.00 | 2615 |
| 1901 2001 | 687S | 44.00 | 4-1.00 | 1.25 | 25.50 | 3.50 | 25.00 | 6.30 | 20.00 | 55.60 | 93.40 | 32.00 | 53.40 | 28.38 | 8.75 | 2950 |
| | 688S | 48.00 | 4-1.00 | 1.25 | 25.50 | 3.50 | 25.00 | 6.30 | 20.00 | 55.60 | 97.40 | 32.00 | 53.40 | 28.38 | 8.75 | 3015 |
| | 689S | 50.00 | 4-1.00 | 1.25 | 25.50 | 3.50 | 25.00 | 6.30 | 20.00 | 55.60 | 100.40 | 32.00 | 53.40 | 29.38 | 8.75 | 3075 |

For reference only. Use certified prints for construction purposes. All dimensions are given in inches.

Approx. weights cover the bedplate only.

*These dimensions are for rough bottom. A finished bedplate is recommended for mounting on structural surfaces. For finished bottom decrease "BC", "BD", "H", & "S" dimensions by .25" with a tolerance of +.00" to −.09" on "BC" dimension.

(Courtesy: PTD, Dresser Industries.)

# 7

# FLEXIBLE POWER TRANSMISSION ELEMENTS

*Flat Belts, V-Belts, and Timing Belts. Chain. Selecting Belts and Chain from Commercial Catalogs.*

Belts and chain are used to connect two or more parallel shafts for the transmission of rotation and power. The shafts may be at a considerable distance from one another, especially in the case of flat belts. Belts also permit a certain amount of twist (see Fig. 7.1).

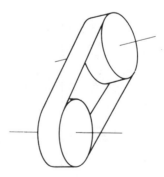

Figure 7.1

All plain belt drives have a constant speed ratio $m_\omega$, except for possible belt slippage, which is defined by

$$m_\omega = \frac{D_{p2}}{D_{p1}} = \frac{\text{rpm}_1}{\text{rpm}_2}$$

in which $D_{p1}$ and $D_{p2}$ are the *pitch diameters* of the pulleys or sheaves (see Fig. 7.2).

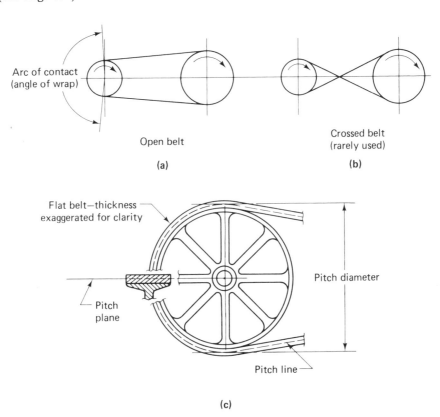

**Figure 7.2**

## 7.1 FLAT BELTS

Flat belts usually run on *crowned* pulleys (see Fig. 7.3). Since flat belts always seek the greatest diameter of the pulley, the crown prevents the belt from leaving the pulley.

Flat belts are made of leather, cotton, various types of rubber and plastic (called *elastomers*) and (rarely) of steel. Steel belts run on un-

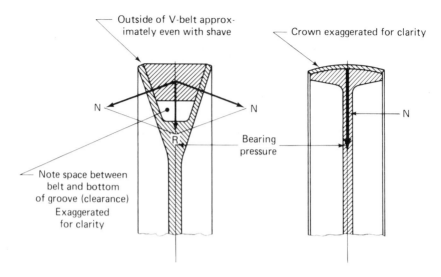

Figure 7.3

crowned, flanged pulleys, the cylindrical surface of which is covered with a thin layer of cork.

*Molded* belts (rubber or plastic) often have an internal reinforcement in the form of a flat core of nylon, polyester, glass fiber, or steel wire.

The *pitch line* of flat belts lies approximately in the central plane parallel to the pulley axis, as shown in Fig. 7.2, or, in the case of reinforced belts in the centroid (the approximate center) of the reinforcement. Since the thickness of flat belts is small compared to the pulley diameter, the latter is often used in speed ratio computations instead of the pitch diameter in the above equation.

To obtain the necessary friction to transmit power, flat belts must be under considerable tension which causes high bearing loads and belt *stretch*, especially if the belts are unreinforced. The stretch causes a particular form of slippage called *creep*, and it makes length computations complicated (see page 94).

Flat belts have excellent flexibility and shock load absorption capability, but all belts slip. Reinforced belts slip less than the unreinforced ones due to the greatly reduced creep. All belt materials except steel are likely to stretch unless they are reinforced, and some belt materials are sensitive to variations in atmospheric humidity. All flat belts require heavy bearings because of the tension requirement. For these reasons, they have been replaced in many instances by the belt types discussed below.

## 7.2  V-BELTS

For the same belt contact area on the pulley or *sheave*, as V-belt pulleys are called, V-belts are narrower but thicker than flat belts and need far less tension for the same normal pressure on the sheave (see Fig. 7.3).

(The normal pressure $N$ on the pulley flanges provides the friction that makes power transmission possible.) V-belts are also quieter, stretch far less (except for fractional horsepower V-belts, all have internal reinforcement), and require less maintenance. Because of their greater thickness, V-belts have more *internal* friction than flat belts, and therefore they have a somewhat lower efficiency (typically 95% vs. 96% to 98%).

V-belts are manufactured in several different types and sizes (Figs. 7.4 & 7.5). Fractional horsepower V-belts are usually not internally reinforced, while industrial sizes (A, B, C, D, and E) all have some form of internal reinforcement (Fig. 7.5).

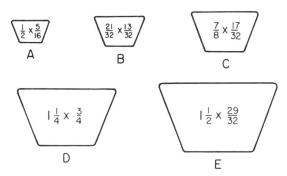

**Figure 7.4**
Reliance Electric Co., Inc., Mishawaka, Indiana

All V-belts have a nylon or other synthetic weave cover for oil and wear resistance and a body of neoprene.

As with reinforced flat belts, the pitch line lies in the centroid of the reinforcement.

Since the portion of the V-belt inside the pitch plane is compressed when the belt is curved around a sheave, it expands in that position so that the angle of the straight sides of the belt is reduced (see Fig. 7.6). This angle (of the straight belt) is standardized at $40°$. For this reason, the groove angle of smaller diameter sheaves must be made correspondingly smaller than $40°$ to prevent binding of the belt. Table 1 in Fig. 7.6 gives the outside diameter (OD) of sheaves, given the pitch diameter, or vice versa, and Table 2 gives recommended groove angles for various pitch diameters and belt sizes.

As stated at the beginning of this chapter, the speed ratio of belts is equal to the inverse ratio of the pitch diameters of the pulleys or sheaves on which they run. Inasmuch as the number of standard sheaves available is necessarily limited, not all desired speed ratios can be obtained and compromises must often be made. For cases in which more accurate speed ratio is important, *adjustable pitch sheaves* are available for single and multiple V-belt drives (see Fig. 7.7). In one type of sheave, one or two flanges are mounted on a screw thread and thus may be turned to move

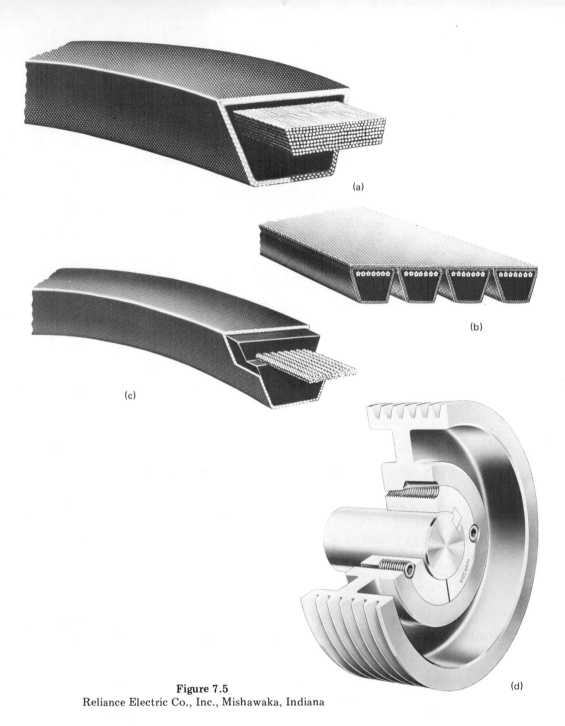

(a)

(b)

(c)

(d)

**Figure 7.5**
Reliance Electric Co., Inc., Mishawaka, Indiana

axially inward or outward. In this manner, the point where the belt rests on the inside of the flanges (and therefore the pitch circle) may be moved radially outward, respectively inward. The adjusted flange is subsequently held in place with a lockscrew. Other designs exist for sheaves having more than two grooves.

## Table 1 Standard V-belt Sections and Sheave Diameters
### (Belt angle, 40 deg)

| Belt section | Belt top width, in. | Belt thickness, in. | Add to pitch diam to get O.D., in. | Min. recommended diam. in. | Range of hp. using 1 or more belts |
|---|---|---|---|---|---|
| A | 1/2 | 1 1/32 | 3/8 | 3.0 | 1/4 to 10 |
| B | 2 1/32 | 7/16 | 1/2 | 5.4 | 1 to 25 |
| C | 7/8 | 1 7/32 | 3/4 | 9.0 | 15 to 100 |
| D | 1 1/4 | 3/4 | 7/8 | 13.0 | 50 to 250 |
| E | 1 1/2 | 1.0 | 1 1/8 | 21.6 | 100 and up |

## Table 2 Recommended Sheave Groove Angles, Degrees
### (Goodrich, 40 deg V belts)

| Sheave O.D., in. | 3/8 in. top | 1/2 in. top | 2 1/32 in. top | 3/4 in. top | 7/8 in. top | 1.0 in. top |
|---|---|---|---|---|---|---|
| 1 1/2 | 26.0 | | | | | |
| 2 | 29.0 | 26.0 | | | | |
| 2 1/2 | 31.6 | 28.5 | 26.0 | | | |
| 3 | 33.7 | 31.0 | 28.5 | 26.0 | | |
| 3 1/2 | 35.0 | 33.0 | 31.0 | 29.0 | 27.0 | |
| 4 | 36.2 | 34.4 | 32.6 | 31.1 | 29.8 | 27.5 |
| 5 | 38.0 | 36.6 | 35.0 | 33.4 | 32.3 | 31.0 |
| 6 | | 38.2 | 36.9 | 35.1 | 33.8 | 32.5 |
| 7 | | | 38.3 | 36.5 | 35.0 | 33.7 |
| 8 | | | | 37.8 | 36.0 | 34.7 |
| 10 | | | | | 37.2 | 35.7 |
| 2 | | | | | | 37.1 |

Values given are for 180 deg arc of contact. For other arcs of contact, correct as follows: 180 to 140 deg. add 1 deg to values in table; 140 to 110 deg, add 2 deg: 110 to 90 deg. add 3 deg.

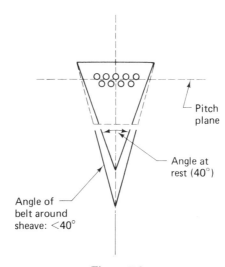

Pitch plane

Angle at rest (40°)

Angle of belt around sheave: <40°

**Figure 7.6**

*Mark's Handbook for Mechanical Engineers*, McGraw Hill.

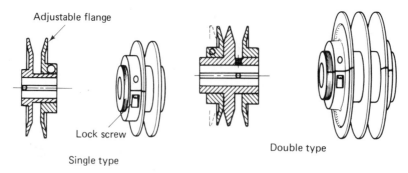

Adjustable flange

Lock screw

Single type

Double type

**Figure 7.7**
Eaton Corporation, Cleveland, Ohio

In cases in which an existing flat wheel of adequate dimensions is already available or in place (for instance, a flywheel), a *V-flat drive* may be economical. Such a drive employs one V-belt pulley with one or more V-belts riding on top of the flat wheel. The rated capacity of a V-flat drive is, of course, lower than that of a regular V-belt drive of the same size, since the wedging action of the V-belt(s) is absent on the flat wheel.

Some types of V-belts lend themselves well to application in certain designs of variable speed drives. Such drives are discussed in Chapter 8.

## 7.3 COMPUTING THE LENGTH OF BELTS

The length of open [Fig. 7.2(a) and (b)] flat and V-belts can be closely approximated by the following equation:

$$L = 2C + 1.57 (D_2 + D_1) + \frac{(D_2 - D_1)^2}{4C}$$

in which $L$ is the pitch length of the belt, $C$ is the center distance, and $D_1$ and $D_2$ are the pitch diameters of the smaller and the larger pulley or sheave. All are in consistent units (inches or metric). It should be noted here that this equation does not provide for belt tensioning requirements (involving stretch), which are of particular importance in unreinforced flat belts. In general, in all belt drives some means of varying the center distance should be provided for tensioning and/or takeup of slack.

According to the Dodge Manufacturing Co. catalog, the *preferred* center distance for V-belts is $(D_2 + 3D_1)/2$ or $D_2$, whichever is greater, Flat belts are joined either by a mechanical connection or are scarfed and cemented together and can thus be made in any desired length. V-belts are mostly molded in one piece and therefore are limited in available sizes. As with speed ratios, compromises in center distance must often be made when off-the-shelf belts are used, as is customary.

Since the center distance is often a restraining factor, it may be necessary to try several different combinations of available belt sizes, number of belts, available standard sheave sizes, and belt lengths to arrive at a satisfactory and economical drive.

## 7.4 SELECTING BELT SECTION, LENGTH, AND PULLEY DIAMETERS

The curves shown in Fig. 7.8 can be used to select a suitable belt section. Input to these curves is the design horsepower (see Chapter 4) and the revolutions per minute of the fastest shaft (usually the driver). For instance, for the transmission of 15 design hp at 1,160 driver rpm, the recommended belt section is B. If a drive consisting of a multiple of single V-belts is contemplated, one may assume a certain number of belts and divide the total design horsepower by this number to arrive at the design load on each belt. It is considered good practice to add a small percentage (from 10% to 15%) to the single belt horsepower rating found. This will compensate for possible load inequalities between belts due to small variations in their length. As mentioned before, the entire process may have to be repeated for different belt and sheave combinations to optimize the drive.

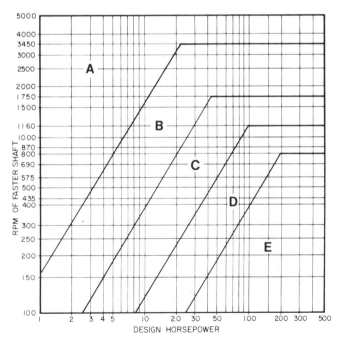

**Figure 7.8**
Reliance Electric Co., Inc., Mishawaka, Indiana

Recommended belt *velocity* is between 1,000 and 5,000 ft/min, with 4,000 ft/min as a good value to start selection of suitable sheave diameters. Remember that in Chapter 4 we mentioned that for a given load in horsepower, the higher the revolutions per minute, the lighter the belt can be. In general, it is therefore economical to stay on the high side of the permissible speed range whenever possible. Belt velocities over 5,000 ft/min should be avoided since the derating influence of centrifugal force becomes excessive around that point. Belt velocity can be found by using the following equation:

$$V_m = \frac{\text{rpm} \times \pi \times D_p}{12}$$

in which $V_m$ is the belt velocity in feet per minute and $D_p$ is the pitch diameter of the sheave in inches.

The precise computation of belt capacity in horsepower is fairly complicated and outside the scope of this text. It involves a basic capacity computation based on the tensile strength of the belt and the subsequent application of several factors pertaining to belt curvature, belt length, and arc-of-contact ratio of the sheaves.

In most cases, actual computation of horsepower rating can be avoided by using the tables provided in manufacturers' catalogs, which also give the steps needed to arrive at the answer. Such tables typically give the rated horsepower for a single belt of a given section based on driver revolutions per minute (usually standard AC motor speeds) and standard pitch diameters of driver and driven sheaves. A combined correction factor for center distance and arc length based on center distance is also given and should be applied to the horsepower rating found earlier.

The following example will clarify how to select V-belts from manufacturers' tables.

Given: A forced draft blower absorbs 24 hp at 750 rpm and runs 24 h/day. It is driven by a single-phase AC motor that runs at 1,750 rpm. The center distance of the drive is to be between 40 in. and 50 in.

Required: Selection of a suitable multiple V-belt drive.

Solution: Find the appropriate service factor in Table 4.1 in Chapter 4. The table gives 1.3. Then design hp = 1.3 × 24 = 31.2 hp. Assume that the drive will use four belts. Applying the 15% additional design horsepower suggested on page 95, you will obtain the design load for each belt from 1.15 × 31.2/4 ≈ 9 hp.

Go to the curves of Fig. 7.8 and enter 9 hp and 1,750 rpm. Find the recommended section A or B (borderline case).

Try the A section first. In the table for A-size belts (Table 7.1) enter the driving motor speed (1,750 in column 2) go downward and find the line with the closest follower speed (751 rpm). Move to the right and find the basic belt horsepower rating in column 9 (1.4 hp). Evidently, this belt is too small.

Now try the B size. Go to the table for B-size belts (Table 7.2) and enter column 1 (1,750 rpm). Then go down to 751 driven rpm. Move horizontally to the right and find the basic horsepower rating of 9.44 in column 8. This is adequate. Continue to the right to find the correction factor for the center distance. Select a center distance of 42.1 (eighth column from right) which is for a B128 belt (top of column) running on driver and driven sheaves of 8.6 in. and 20 in. pitch diameter (columns 5 and 6). Now read the correction factor, which is 1.05 (below 42.1 in the same column). The corrected belt rating is 1.05 × 9.44 = 9.9 hp.

The belt speed is

$$\frac{1,750 \times 8.6 \times \pi}{12} = 3,940 \text{ ft/min} \qquad (\text{almost } 4000 - \text{very good})$$

## 7.5 CHAIN DRIVES

Chain drives may be thought to have been developed from belt drives in an effort to overcome the drawback of slippage, in analogy with the imagined development of gearing from friction drives (see Chapter 5).

There are many different types of chain. The most common are *roller chain* and *inverted tooth ("silent") chain*. Some types of roller chain are illustrated in Fig. 7.9.

Roller chain consists essentially of a great number of highly loaded, small precision journal bearings, which implies that oil lubrication is essential.

Roller chain link plates (see Fig. 7.10) may be modified to provide attachment points which makes possible their use in industrial conveying systems.

Several methods of oil lubrication are in use (grease is less suitable). Some types of roller chain have bushings made from sintered steel which are charged with oil. Such chain is self-lubricating. (See discussion of sintered bearings in Chap. 9.)

Chain runs on *sprockets*, the pitch diameter of which is measured from the center of a nested roller to the center of the one diametrically opposite. In contrast with gearing, the pitch is the *straight* distance between the centers of adjacent rollers and is *not* measured along the pitch circle (see Fig. 7.11).

The speed ratio for chain sprockets is the same as that for gears, namely, that of the number of teeth: $N_2/N_1$. However, both sprockets

# TABLE 7.1

## A  STOCK DRIVE SELECTIONS

Driven Speeds and HP per Belt @ 1.0 Service Factor†

| Speed Ratio | Pitch Diam. Driver | Pitch Diam. Driven | 3450 RPM Driver — Driven RPM | 3450 HP/Belt Stl. Cbl. | 3450 HP/Belt Std. Life II | 3450 HP/Belt DYNA-COG | 1750 RPM Driver — Driven RPM | 1750 HP/Belt Stl. Cbl. | 1750 HP/Belt Std. Life II | 1750 HP/Belt DYNA-COG | 1160 RPM Driver — Driven RPM | 1160 HP/Belt Stl. Cbl. | 1160 HP/Belt Std. Life II | 1160 HP/Belt DYNA-COG | A26 | A31 | A33 | A35 | A38 |
|---|---|---|---|---|---|---|---|---|---|---|---|---|---|---|---|---|---|---|---|
| 1.24 | 3.4 | 4.2 | 2782 | 3.48 | 3.88 | 4.73 | 1411 | 2.17 | 2.50 | 3.17 | 935 | 1.57 | 1.84 | 2.41 | 7.7 | 10.2 | 11.2 | 12.2 | 13.7 |
| 1.24 | 4.2 | 5.2 | 2782 | 4.86 | 5.41 | 6.69 | 1411 | 3.05 | 3.48 | 4.41 | 935 | 2.20 | 2.54 | 3.30 | 6.3 | 8.8 | 9.8 | 10.8 | 12.3 |
| 1.25 | 3.2 | 4.0 | 2760 | 3.12 | 3.48 | 4.22 | 1400 | 1.95 | 2.25 | 2.86 | 928 | 1.41 | 1.67 | 2.18 | 8.0 | 10.5 | 11.5 | 12.5 | 14.0 |
| 1.25 | 4.0 | 5.0 | 2760 | 4.54 | 5.06 | 6.23 | 1400 | 2.84 | 3.24 | 4.11 | 928 | 2.04 | 2.37 | 3.08 | 6.6 | 9.1 | 10.1 | 11.1 | 12.6 |
| 1.25 | 4.8 | 6.0 | 2760 | 5.76 | 6.43 | 8.02 | 1400 | 3.70 | 4.19 | 5.30 | 928 | 2.66 | 3.06 | 3.94 |  | 7.6 | 8.7 | 9.7 | 11.2 |
| Arc-Length Correction Factor → |  |  |  |  |  |  |  |  |  |  |  |  |  |  | .79 | .83 | .85 | .86 | .87 |
| 1.25 | 5.6 | 7.0 | 2760 | 6.75 | 7.56 | 9.59 | 1400 | 4.51 | 5.09 | 6.43 | 928 | 3.25 | 3.72 | 4.78 |  |  | 7.2 | 8.2 | 9.7 |
| 1.26 | 3.8 | 4.8 | 2738 | 4.21 | 4.69 | 5.76 | 1389 | 2.63 | 3.00 | 3.81 | 921 | 1.89 | 2.20 | 2.86 | 6.9 | 9.4 | 10.4 | 11.4 | 12.9 |
| 1.27 | 3.0 | 3.8 | 2717 | 2.75 | 3.07 | 3.69 | 1378 | 1.72 | 2.00 | 2.54 | 913 | 1.25 | 1.49 | 1.96 | 8.3 | 10.8 | 11.8 | 12.8 | 14.3 |
| 1.28 | 3.6 | 4.6 | 2695 | 3.87 | 4.31 | 5.27 | 1367 | 2.41 | 2.76 | 3.50 | 906 | 1.74 | 2.03 | 2.64 | 7.2 | 9.7 | 10.7 | 11.7 | 13.2 |
| 1.28 | 5.0 | 6.4 | 2695 | 6.05 | 6.75 | 8.46 | 1367 | 3.92 | 4.43 | 5.60 | 906 | 2.81 | 3.23 | 4.16 |  | 7.2 | 8.2 | 9.2 | 10.7 |
| Arc-Length Correction Factor → |  |  |  |  |  |  |  |  |  |  |  |  |  |  | .80 | .83 | .84 | .86 | .87 |
| 1.28 | 6.4 | 8.2 | 2695 | 7.51 | 8.47 | 10.92 | 1367 | 5.30 | 5.96 | 7.53 | 906 | 3.84 | 4.37 | 5.59 |  |  |  |  |  |
| 1.29 | 7.0 | 9.0 |  |  |  |  | 1357 | 5.86 | 6.58 | 8.31 | 899 | 4.27 | 4.84 | 6.19 |  |  |  |  |  |
| 1.30 | 4.0 | 5.2 | 2654 | 4.57 | 5.09 | 6.26 | 1346 | 2.86 | 3.26 | 4.12 | 892 | 2.05 | 2.38 | 3.09 | 6.4 | 8.9 | 9.9 | 10.9 | 12.4 |
| 1.30 | 4.6 | 6.0 | 2654 | 5.51 | 6.14 | 7.63 | 1346 | 3.50 | 3.97 | 5.02 | 892 | 2.52 | 2.90 | 3.74 |  | 7.8 | 8.8 | 9.8 | 11.3 |
| 1.31 | 3.2 | 4.2 | 2634 | 3.16 | 3.52 | 4.26 | 1336 | 1.92 | 2.27 | 2.88 | 885 | 1.43 | 1.68 | 2.20 | 7.8 | 10.3 | 11.3 | 12.3 | 13.8 |
| Arc-Length Correction Factor → |  |  |  |  |  |  |  |  |  |  |  |  |  |  | .79 | .83 | .84 | .86 | .87 |
| 1.32 | 3.8 | 5.0 | 2614 | 4.25 | 4.72 | 5.79 | 1326 | 2.64 | 3.02 | 3.82 | 879 | 1.90 | 2.21 | 2.87 | 6.7 | 9.2 | 10.2 | 11.2 | 12.7 |
| 1.33 | 3.0 | 4.0 | 2594 | 2.79 | 3.11 | 3.73 | 1316 | 1.74 | 2.02 | 2.56 | 872 | 1.27 | 1.50 | 1.97 | 8.1 | 10.6 | 11.6 | 12.6 | 14.2 |
| 1.33 | 3.6 | 4.8 | 2594 | 3.91 | 4.35 | 5.30 | 1316 | 2.43 | 2.78 | 3.52 | 872 | 1.75 | 2.04 | 2.65 | 7.0 | 9.5 | 10.5 | 11.5 | 13.0 |
| 1.33 | 4.2 | 5.6 | 2594 | 4.92 | 5.47 | 6.75 | 1316 | 3.09 | 3.51 | 4.44 | 872 | 2.22 | 2.56 | 3.32 | 5.9 | 8.4 | 9.4 | 10.4 | 11.9 |
| 1.33 | 4.8 | 6.4 | 2594 | 5.81 | 6.48 | 8.08 | 1316 | 3.72 | 4.22 | 5.32 | 872 | 2.67 | 3.08 | 3.96 |  | 7.3 | 8.3 | 9.3 | 10.8 |
| Arc-Length Correction Factor → |  |  |  |  |  |  |  |  |  |  |  |  |  |  | .79 | .83 | .84 | .86 | .87 |
| 1.35 | 3.4 | 4.6 | 2556 | 3.56 | 3.96 | 4.80 | 1296 | 2.21 | 2.53 | 3.21 | 859 | 1.59 | 1.87 | 2.43 | 7.3 | 9.9 | 10.9 | 11.9 | 13.4 |
| 1.35 | 5.2 | 7.0 | 2556 | 6.35 | 7.09 | 8.90 | 1296 | 4.14 | 4.68 | 5.90 | 859 | 2.98 | 3.41 | 4.38 |  | 7.5 |  | 8.5 | 10.0 |
| 1.37 | 3.8 | 5.2 | 2518 | 4.27 | 4.75 | 5.82 | 1277 | 2.66 | 3.03 | 3.84 | 847 | 1.91 | 2.22 | 2.88 | 6.5 | 9.1 | 10.1 | 11.1 | 12.6 |
| 1.37 | 6.0 | 8.2 | 2518 | 7.22 | 8.11 | 10.35 | 1277 | 4.94 | 5.56 | 7.02 | 847 | 3.57 | 4.07 | 5.21 |  |  |  |  | 8.4 |
| 1.39 | 3.6 | 5.0 | 2482 | 3.93 | 4.37 | 5.33 | 1259 | 2.44 | 2.79 | 3.53 | 835 | 1.76 | 2.05 | 2.66 | 6.9 | 9.4 | 10.4 | 11.4 | 12.9 |
| Arc-Length Correction Factor → |  |  |  |  |  |  |  |  |  |  |  |  |  |  | .79 | .83 | .84 | .85 | .87 |
| 1.39 | 4.6 | 6.4 | 2482 | 5.56 | 6.18 | 7.68 | 1259 | 3.53 | 4.00 | 5.05 | 835 | 2.53 | 2.92 | 3.76 |  | 7.5 | 8.5 | 9.5 | 11.0 |
| 1.40 | 3.0 | 4.2 | 2464 | 2.82 | 3.14 | 3.77 | 1250 | 1.76 | 2.04 | 2.58 | 829 | 1.28 | 1.51 | 1.98 | 8.0 | 10.5 | 11.5 | 12.5 | 14.0 |
| 1.40 | 4.0 | 5.6 | 2464 | 4.63 | 5.14 | 6.31 | 1250 | 2.88 | 3.28 | 4.15 | 829 | 2.07 | 2.40 | 3.11 | 6.1 | 8.6 | 9.6 | 10.6 | 12.1 |
| 1.40 | 5.0 | 7.0 | 2464 | 6.11 | 6.82 | 8.52 | 1250 | 3.95 | 4.46 | 5.63 | 829 | 2.84 | 3.25 | 4.18 |  |  | 7.7 | 8.7 | 10.2 |
| 1.41 | 3.4 | 4.8 | 2447 | 3.58 | 3.98 | 4.83 | 1241 | 2.22 | 2.55 | 3.22 | 823 | 1.60 | 1.87 | 2.44 | 7.2 | 9.7 | 10.7 | 11.7 | 13.2 |
| Arc-Length Correction Factor → |  |  |  |  |  |  |  |  |  |  |  |  |  |  | .79 | .83 | .84 | .85 | .87 |

1-6 Groove Stock Sheaves

V-belt Number and Approx. Center Distance* (Con't. next page)

## TABLE 7.1 (cont.)

*Arc-Length Correction Factor →*

**Group 1** (ratios 1.41–1.46)

| Ratio | | | | | | | | | | | | | | ALCF | .79 | .82 | .84 | .85 | .87 |
|---|---|---|---|---|---|---|---|---|---|---|---|---|---|---|---|---|---|---|---|
| 1.41 | 6.4 | 9.0 | 2447 | 7.57 | 8.54 | 10.99 | 1241 | 5.33 | 5.99 | 7.56 | 823 | 3.86 | 4.39 | 5.62 |  |  |  |  |  |
| 1.43 | 4.2 | 6.0 | 2413 | 4.96 | 5.51 | 6.79 | 1224 | 3.11 | 3.53 | 4.46 | 811 | 2.23 | 2.58 | 3.33 |  | 8.1 | 9.1 | 10.1 | 11.6 |
| 1.44 | 3.2 | 4.6 | 2396 | 3.22 | 3.58 | 4.32 | 1215 | 2.00 | 2.30 | 2.91 | 806 | 1.44 | 1.70 | 2.21 |  | 10.0 | 11.0 | 12.0 | 13.5 |
| 1.44 | 3.6 | 5.2 | 2396 | 3.96 | 4.39 | 5.35 | 1215 | 2.45 | 2.80 | 3.54 | 806 | 1.76 | 2.06 | 2.67 | 7.5 | 9.2 | 10.2 | 11.2 | 12.7 |
| 1.46 | 4.8 | 7.0 | 2363 | 5.87 | 6.53 | 8.13 | 1199 | 3.75 | 4.24 | 5.35 | 795 | 2.69 | 3.09 | 3.98 | 6.7 |  | 7.8 | 8.8 | 10.3 |

**Group 2** (ratios 1.46–1.50)

| Ratio | | | | | | | | | | | | | | ALCF | .78 | .82 | .83 | .85 | .86 |
|---|---|---|---|---|---|---|---|---|---|---|---|---|---|---|---|---|---|---|---|---|
| 1.46 | 5.6 | 8.2 | 2363 | 6.86 | 7.67 | 9.70 | 1199 | 4.57 | 5.15 | 6.49 | 795 | 3.29 | 3.76 | 4.81 |  |  |  |  | 8.7 |
| 1.47 | 3.4 | 5.0 | 2347 | 3.60 | 4.00 | 4.85 | 1190 | 2.23 | 2.56 | 3.23 | 789 | 1.61 | 1.88 | 2.45 | 7.0 | 9.5 | 10.5 | 11.5 | 13.0 |
| 1.47 | 3.8 | 5.6 | 2347 | 4.31 | 4.79 | 5.86 | 1190 | 2.68 | 3.05 | 3.86 | 789 | 1.92 | 2.24 | 2.90 | 6.2 | 8.7 | 9.7 | 10.7 | 12.2 |
| 1.50 | 3.2 | 4.8 | 2300 | 3.24 | 3.60 | 4.34 | 1167 | 2.01 | 2.31 | 2.92 | 773 | 1.45 | 1.71 | 2.22 | 7.3 | 9.8 | 10.8 | 11.9 | 13.4 |
| 1.50 | 4.0 | 6.0 | 2300 | 4.66 | 5.17 | 6.35 | 1167 | 2.90 | 3.30 | 4.17 | 773 | 2.08 | 2.41 | 3.12 | 8.2 | 9.2 | 10.3 | 11.8 |  |

**Group 3** (ratios 1.50–1.53)

| Ratio | | | | | | | | | | | | | | ALCF | .78 | .82 | .84 | .85 | .86 |
|---|---|---|---|---|---|---|---|---|---|---|---|---|---|---|---|---|---|---|---|---|
| 1.50 | 6.0 | 9.0 | 2300 | 7.27 | 8.16 | 10.40 | 1167 | 4.97 | 5.59 | 7.04 | 773 | 3.59 | 4.08 | 5.23 |  |  |  |  |  |
| 1.51 | 7.0 | 10.6 |  |  |  |  | 1159 | 5.90 | 6.62 | 8.36 | 768 | 4.30 | 4.88 | 6.22 |  |  |  |  |  |
| 1.52 | 4.2 | 6.4 | 2270 | 4.99 | 5.54 | 6.82 | 1151 | 3.12 | 3.54 | 4.47 | 763 | 2.24 | 2.59 | 3.34 | 7.8 | 8.8 | 9.8 | 11.3 |  |
| 1.52 | 4.6 | 7.0 | 2270 | 5.60 | 6.23 | 7.72 | 1151 | 3.55 | 4.02 | 5.07 | 763 | 2.55 | 2.93 | 3.77 | 6.9 | 8.0 | 9.0 | 10.5 |  |
| 1.53 | 3.0 | 4.6 | 2255 | 2.87 | 3.18 | 3.81 | 1144 | 1.78 | 2.06 | 2.60 | 758 | 1.29 | 1.53 | 1.99 | 7.6 | 10.2 | 11.2 | 12.2 | 13.7 |

**Group 4** (ratios 1.53–1.58)

| Ratio | | | | | | | | | | | | | | ALCF | .78 | .82 | .83 | .85 | .86 |
|---|---|---|---|---|---|---|---|---|---|---|---|---|---|---|---|---|---|---|---|---|
| 1.53 | 3.4 | 5.2 | 2255 | 3.62 | 4.02 | 4.87 | 1144 | 2.24 | 2.56 | 3.24 | 758 | 1.61 | 1.89 | 2.45 | 6.8 | 9.4 | 10.4 | 11.4 | 12.9 |
| 1.56 | 3.2 | 5.0 | 2212 | 3.26 | 3.61 | 4.35 | 1122 | 2.01 | 2.32 | 2.93 | 744 | 1.46 | 1.71 | 2.23 | 7.2 | 9.7 | 10.7 | 11.5 | 13.2 |
| 1.56 | 3.6 | 5.6 | 2212 | 3.99 | 4.43 | 5.38 | 1122 | 2.46 | 2.82 | 3.56 | 744 | 1.77 | 2.07 | 2.68 | 6.3 | 8.9 | 9.9 | 10.9 | 12.4 |
| 1.58 | 3.8 | 6.0 | 2184 | 4.34 | 4.82 | 5.88 | 1108 | 2.69 | 3.07 | 3.87 | 734 | 1.93 | 2.24 | 2.90 | 5.9 | 8.4 | 9.4 | 10.4 | 11.9 |
| 1.58 | 5.2 | 8.2 | 2184 | 6.42 | 7.16 | 8.97 | 1108 | 4.18 | 4.71 | 5.94 | 734 | 3.00 | 3.44 | 4.41 |  |  |  |  | 9.0 |

**Group 5** (ratios 1.60–1.64)

| Ratio | | | | | | | | | | | | | | ALCF | .78 | .82 | .83 | .85 | .86 |
|---|---|---|---|---|---|---|---|---|---|---|---|---|---|---|---|---|---|---|---|---|
| 1.60 | 3.0 | 4.8 | 2156 | 2.88 | 3.20 | 3.82 | 1094 | 1.79 | 2.06 | 2.61 | 725 | 1.30 | 1.53 | 2.00 | 7.5 | 10.0 | 11.0 | 12.0 | 13.5 |
| 1.60 | 4.0 | 6.4 | 2156 | 4.68 | 5.20 | 6.37 | 1094 | 2.91 | 3.31 | 4.18 | 725 | 2.09 | 2.42 | 3.13 | 7.0 | 7.9 | 8.9 | 9.9 | 11.4 |
| 1.61 | 5.6 | 9.0 | 2143 | 6.89 | 7.71 | 9.73 | 1087 | 4.59 | 5.16 | 6.51 | 720 | 3.30 | 3.77 | 4.83 |  | 9.5 | 10.5 | 11.5 | 13.0 |
| 1.63 | 3.2 | 5.2 | 2117 | 3.27 | 3.63 | 4.36 | 1074 | 2.02 | 2.23 | 2.93 | 712 | 1.46 | 1.71 | 2.23 |  |  |  | 7.6 | 9.2 |
| 1.64 | 5.0 | 8.2 | 2104 | 6.18 | 6.88 | 8.59 | 1067 | 3.98 | 4.49 | 5.66 | 707 | 2.86 | 3.28 | 4.20 |  |  |  |  |  |

**Note**—Interpolate center distance for belts not listed. See listings on pages 33-50 thru 33-56.

**To Find Number of Belts**—Design hp = Normal running hp or motor nameplate rating × proper service factor from page 33-5. Corrected HP/Belt = HP/Belt value, from table above, × Arc and Length Correction Factor (located below center distance selected). Add 0.10 to factor if shown in shaded area. Number of Belts = Design HP ÷ Corrected HP/Belt.

‡ Includes allowance for speed ratio.

▲ For sheaves with other numbers of grooves, see Made-to-Order type.

\* Provisions should be made for belt installation and take-up. Refer to Table 10, page 33-34.

(Courtesy: PTD, Dresser Industries.)

# TABLE 7.2

# A STOCK DRIVE SELECTIONS

Driven Speeds and HP per Belt @ 1.0 Service Factor†

V-belt Number and Approx. Center Distance* (Con't. next page)

| Speed Ratio | Pitch Diam. Driver | Pitch Diam. Driven | 3450 RPM Driver — Driven RPM | 3450 HP/Belt Stl. Cbl. | 3450 HP/Belt Sld. Life II | 3450 HP/Belt DYNA-COG | 1750 RPM Driver — Driven RPM | 1750 HP/Belt Stl. Cbl. | 1750 HP/Belt Sld. Life II | 1750 HP/Belt DYNA-COG | 1160 RPM Driver — Driven RPM | 1160 HP/Belt Stl. Cbl. | 1160 HP/Belt Sld. Life II | 1160 HP/Belt DYNA-COG | A31 | A33 | A35 | A38 |
|---|---|---|---|---|---|---|---|---|---|---|---|---|---|---|---|---|---|---|
| 2.50 | 3.6 | 9.0 | 1380 | 4.06 | 4.50 | 5.45 | 700 | 2.50 | 3.84 | 3.59 | 464 | 1.80 | 2.09 | 2.70 | … | … | … | 9.4 |
| 2.50 | 4.8 | 12.0 | 1380 | 5.96 | 6.63 | 8.23 | 700 | 3.80 | 5.71 | 5.40 | 464 | 2.72 | 3.12 | 4.01 | … | … | 7.8 | … |
| 2.50 | 6.0 | 15.0 | 1380 | 7.35 | 8.24 | 10.48 | 700 | 5.01 | 7.09 | 7.08 | 464 | 3.61 | 4.11 | 5.25 | … | … | … | … |
| 2.52 | 4.2 | 10.6 | 1369 | 5.07 | 5.62 | 6.90 | 694 | 3.16 | 4.83 | 4.51 | 460 | 2.27 | 2.61 | 3.37 | … | … | … | … |
| 2.56 | 3.2 | 8.2 | 1348 | 3.33 | 3.68 | 4.42 | 684 | 2.05 | 3.13 | 2.96 | 453 | 1.48 | 1.73 | 2.25 | 6.7 | 7.8 | 8.8 | 10.4 |
| Arc-Length Correction Factor → | | | | | | | | | | | | | | | .74 | .77 | .79 | .81 |
| 2.57 | 7.0 | 18.0 | — | — | — | — | 681 | 5.95 | | 8.40 | 451 | 4.33 | 4.90 | 6.25 | … | … | … | … |
| 2.61 | 4.6 | 12.0 | 1322 | 5.68 | 6.31 | 7.80 | 670 | 3.59 | 5.43 | 5.11 | 444 | 2.57 | 2.96 | 3.80 | … | … | … | 9.5 |
| 2.65 | 3.4 | 9.0 | 1302 | 3.70 | 4.10 | 4.94 | 660 | 2.28 | 3.49 | 3.28 | 438 | 1.64 | 1.91 | 2.48 | … | … | … | … |
| 2.65 | 4.0 | 10.6 | 1302 | 4.75 | 5.26 | 6.43 | 660 | 2.94 | 4.51 | 4.21 | 438 | 2.11 | 2.44 | 3.15 | … | … | 7.9 | … |
| 2.68 | 5.6 | 15.0 | 1287 | 6.96 | 7.77 | 9.79 | 653 | 4.62 | 6.69 | 6.54 | 433 | 3.32 | 3.79 | 4.85 | … | … | … | … |
| Arc-Length Correction Factor → | | | | | | | | | | | | | | | .74 | .77 | .79 | .81 |
| 2.73 | 3.0 | 8.2 | 1264 | 2.95 | 3.26 | 3.89 | 641 | 1.82 | 2.75 | 2.64 | 425 | 1.32 | 1.55 | 2.02 | 6.9 | 7.9 | 9.0 | 10.5 |
| 2.79 | 3.8 | 10.6 | 1237 | 4.41 | 4.89 | 5.95 | 627 | 2.72 | 4.19 | 3.90 | 416 | 1.96 | 2.27 | 2.93 | … | … | … | … |
| 2.81 | 3.2 | 9.0 | 1228 | 3.33 | 3.69 | 4.42 | 623 | 2.05 | 3.13 | 2.96 | 413 | 1.48 | 1.73 | 2.25 | … | … | 8.1 | 9.6 |
| 2.81 | 6.4 | 18.0 | 1228 | 7.69 | 8.66 | 11.11 | 623 | 5.39 | 7.43 | 7.62 | 413 | 3.90 | 4.43 | 5.66 | … | … | … | … |
| 2.86 | 4.2 | 12.0 | 1206 | 5.07 | 5.63 | 6.91 | 612 | 3.16 | 4.84 | 4.51 | 406 | 2.27 | 2.62 | 3.37 | … | … | … | … |
| Arc-Length Correction Factor → | | | | | | | | | | | | | | | .74 | .77 | .77 | .80 |
| 2.88 | 5.2 | 15.0 | 1198 | 6.49 | 7.23 | 9.04 | 608 | 4.21 | 6.23 | 5.98 | 403 | 3.03 | 3.46 | 4.43 | … | … | … | … |
| 2.94 | 3.6 | 10.6 | 1173 | 4.06 | 4.50 | 5.46 | 595 | 2.50 | 3.85 | 3.60 | 395 | 1.80 | 2.09 | 2.70 | … | … | … | … |
| 3.00 | 3.0 | 9.0 | 1150 | 2.95 | 3.27 | 3.89 | 583 | 1.82 | 2.76 | 2.64 | 387 | 1.32 | 1.55 | 2.02 | … | … | … | … |
| 3.00 | 4.0 | 12.0 | 1150 | 4.75 | 5.26 | 6.44 | 583 | 2.95 | 4.52 | 4.21 | 387 | 2.11 | 2.44 | 3.15 | … | … | 8.2 | … |
| 3.00 | 5.0 | 15.0 | 1150 | 6.24 | 6.94 | 8.65 | 583 | 4.01 | 5.98 | 5.69 | 387 | 2.88 | 3.29 | 4.22 | … | … | … | 9.8 |
| Arc-Length Correction Factor → | | | | | | | | | | | | | | | .74 | .77 | .78 | .81 |
| 3.00 | 6.0 | 18.0 | 1150 | 7.36 | 8.25 | 10.49 | 583 | 5.01 | 7.09 | 7.09 | 387 | 3.62 | 4.11 | 5.26 | … | … | … | … |
| 3.13 | 3.4 | 10.6 | 1106 | 3.70 | 4.10 | 4.95 | 561 | 2.28 | 2.61 | 3.28 | 372 | 1.64 | 1.91 | 2.48 | … | … | … | … |
| 3.13 | 4.8 | 15.0 | 1102 | 5.97 | 6.63 | 8.23 | 559 | 3.80 | 4.29 | 5.40 | 371 | 2.73 | 3.13 | 4.01 | … | … | 8.2 | … |
| 3.16 | 3.8 | 12.0 | 1092 | 4.41 | 4.89 | 5.96 | 554 | 2.73 | 3.10 | 3.91 | 367 | 1.96 | 2.27 | 2.93 | … | … | … | 9.8 |
| 3.21 | 5.6 | 18.0 | 1075 | 6.96 | 7.77 | 9.80 | 545 | 4.62 | 5.20 | 6.54 | 361 | 3.32 | 3.79 | 4.85 | … | … | … | … |
| Arc-Length Correction Factor → | | | | | | | | | | | | | | | .74 | .77 | .77 | .80 |

# TABLE 7.2 (cont.)

## Arc-Length Correction Factor →

| | | | | | | | | | | | | | | | |
|---|---|---|---|---|---|---|---|---|---|---|---|---|---|---|---|
| 3.26 | 4.6 | 15.0 | 1058 | 5.69 | 6.31 | 7.81 | 537 | 3.59 | 4.06 | 5.11 | 356 | 2.57 | 2.96 | 3.80 | 8.0 |
| 3.31 | 3.2 | 10.6 | 1042 | 3.33 | 3.69 | 4.43 | 529 | 2.05 | 2.35 | 2.97 | 350 | 1.48 | 1.74 | 2.25 | |
| 3.33 | 3.6 | 12.0 | 1036 | 4.07 | 4.50 | 5.46 | 526 | 2.50 | 2.86 | 3.60 | 348 | 1.80 | 2.09 | 2.71 | |
| 3.46 | 5.2 | 18.0 | 997 | 6.50 | 7.24 | 9.05 | 506 | 4.22 | 4.75 | 5.98 | 335 | 3.03 | 3.46 | 4.43 | |
| 3.53 | 3.0 | 10.6 | 977 | 2.95 | 3.27 | 3.89 | 496 | 1.82 | 2.10 | 2.64 | 329 | 1.32 | 1.56 | 2.02 | 8.1 |

## Arc-Length Correction Factor →   .77

| | | | | | | | | | | | | | | | |
|---|---|---|---|---|---|---|---|---|---|---|---|---|---|---|---|
| 3.53 | 3.4 | 12.0 | 977 | 3.71 | 4.10 | 4.95 | 496 | 2.28 | 2.61 | 3.28 | 329 | 1.64 | 1.92 | 2.48 | |
| 3.57 | 4.2 | 15.0 | 966 | 5.08 | 5.63 | 6.91 | 490 | 3.16 | 3.59 | 4.52 | 325 | 2.27 | 2.62 | 3.37 | |
| 3.60 | 5.0 | 18.0 | 958 | 6.24 | 6.94 | 8.65 | 486 | 4.01 | 4.52 | 5.69 | 322 | 2.88 | 3.30 | 4.23 | |
| 3.75 | 3.2 | 12.0 | 920 | 3.33 | 3.69 | 4.43 | 467 | 2.05 | 2.35 | 2.97 | 309 | 1.48 | 1.74 | 2.25 | |
| 3.75 | 4.0 | 15.0 | 920 | 4.75 | 5.27 | 6.44 | 467 | 2.95 | 3.35 | 4.21 | 309 | 2.11 | 2.44 | 3.15 | |

## Arc-Length Correction Factor →

| | | | | | | | | | | | | | | |
|---|---|---|---|---|---|---|---|---|---|---|---|---|---|---|
| 3.75 | 4.8 | 18.0 | 920 | 5.97 | 6.64 | 8.24 | 467 | 3.80 | 4.29 | 5.40 | 309 | 2.73 | 3.13 | 4.01 |
| 3.91 | 4.6 | 18.0 | 882 | 5.69 | 6.32 | 7.81 | 448 | 3.59 | 4.06 | 5.11 | 297 | 2.58 | 2.96 | 3.80 |
| 3.95 | 3.8 | 15.0 | 873 | 4.42 | 4.89 | 5.96 | 443 | 2.73 | 3.10 | 3.91 | 294 | 1.96 | 2.27 | 2.93 |
| 4.00 | 3.0 | 12.0 | 863 | 2.95 | 3.27 | 3.89 | 438 | 1.82 | 2.10 | 2.64 | 290 | 1.32 | 1.56 | 2.02 |
| 4.17 | 3.6 | 15.0 | 827 | 4.07 | 4.51 | 5.46 | 420 | 2.51 | 2.86 | 3.60 | 278 | 1.80 | 2.09 | 2.71 |

## Arc-Length Correction Factor →

| | | | | | | | | | | | | | | |
|---|---|---|---|---|---|---|---|---|---|---|---|---|---|---|
| 4.29 | 4.2 | 18.0 | 804 | 5.08 | 5.63 | 6.91 | 408 | 3.17 | 3.59 | 4.52 | 270 | 2.27 | 2.62 | 3.37 |
| 4.41 | 3.4 | 15.0 | 782 | 3.71 | 4.11 | 4.95 | 397 | 2.28 | 2.61 | 3.28 | 263 | 1.64 | 1.92 | 2.48 |
| 4.50 | 4.0 | 18.0 | 767 | 4.75 | 5.27 | 6.44 | 389 | 2.95 | 3.35 | 4.21 | 258 | 2.11 | 2.44 | 3.15 |
| 4.69 | 3.2 | 15.0 | 736 | 3.34 | 3.70 | 4.43 | 373 | 2.05 | 2.36 | 2.97 | 247 | 1.48 | 1.74 | 2.25 |
| 4.74 | 3.8 | 18.0 | 728 | 4.42 | 4.89 | 5.96 | 369 | 2.73 | 3.10 | 3.91 | 245 | 1.96 | 2.27 | 2.93 |

## Arc-Length Correction Factor →

| | | | | | | | | | | | | | | |
|---|---|---|---|---|---|---|---|---|---|---|---|---|---|---|
| 5.00 | 3.0 | 15.0 | 690 | 2.95 | 3.27 | 3.90 | 350 | 1.83 | 2.10 | 2.65 | 232 | 1.32 | 1.56 | 2.02 |
| 5.00 | 3.6 | 18.0 | 690 | 4.07 | 4.51 | 5.46 | 350 | 2.51 | 2.86 | 3.60 | 232 | 1.80 | 2.09 | 2.71 |
| 5.29 | 3.4 | 18.0 | 652 | 3.71 | 4.11 | 4.95 | 331 | 2.28 | 2.61 | 3.28 | 219 | 1.64 | 1.92 | 2.48 |
| 5.63 | 3.2 | 18.0 | 613 | 3.34 | 3.70 | 4.43 | 311 | 2.05 | 2.36 | 2.97 | 206 | 1.48 | 1.74 | 2.25 |
| 6.00 | 3.0 | 18.0 | 575 | 2.96 | 3.27 | 3.90 | 292 | 1.83 | 2.10 | 2.65 | 193 | 1.32 | 1.56 | 2.02 |

**Note**—Interpolate center distance for belts not listed. See listings on pages 33-50 thru 33-56.

**To Find Number of Belts**—Design hp = Normal running hp or motor nameplate rating × proper service factor from page 33-5. Corrected HP/Belt = HP/Belt value, from table above, × Arc and Length Correction Factor (located below center distance selected). Add 0.10 to factor if shown in shaded area. Number of Belts = Design HP ÷ Corrected HP/Belt.

† Includes allowance for speed ratio.
▲ For sheaves with other numbers of grooves, see Made-to-Order type.
* Provisions should be made for belt installation and take-up. Refer to Table 10, page 33-34.

(Courtesy: PTD, Dresser Industries.)

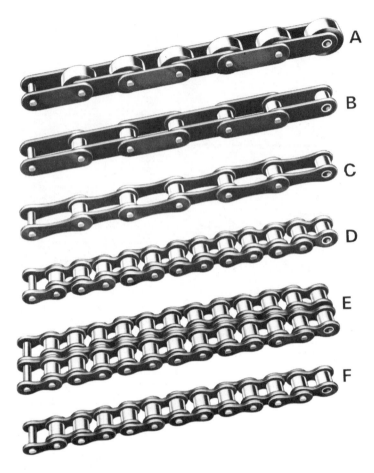

**Figure 7.9**
Reliance Electric Co., Inc., Mishawaka, Indiana

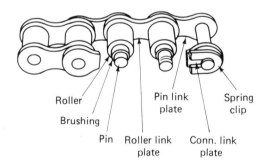

Roller

Brushing

Pin  Roller link
       plate

Pin link
plate

Conn. link
plate

Spring
clip

**Figure 7.10**
Boston Gear, Quincy, Mass

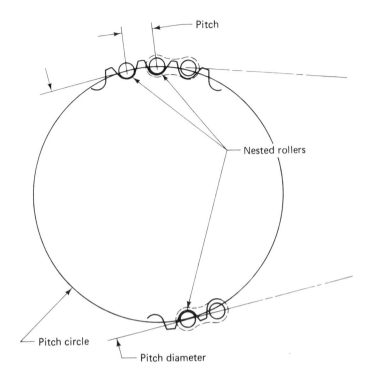

Figure 7.11

run in the same direction. Speed ratios over 7 are generally not recommended for chain drives.

The chain length must, of course, be a multiple of the pitch, i.e., the length of one shackle. It is preferable that the number of pitches be *even* so that the need for a special offset link is avoided. A drawback of roller chain drives is the so-called *chordal effect* (see Fig. 7.12) which causes slight variations in the speed ratio. The inconstancy is caused by the slightly varying effective pitch *radii* of chain positions when moving from one tooth to the next at the point where the chain engages the driver sprocket.

The chordal effect is especially noticeable in sprockets having low tooth numbers. For this reason, the number of teeth of the smaller sprocket should not be less than 12 for slow speeds, 17 for medium speeds, and 21 for high speeds.

Maximum permissible chain speed varies widely and depends on the pitch, the method of lubrication used, and the number of teeth of the smaller sprocket. Manufacturers' recommendations should be followed.

The center distance should be between 30 and 50 pitches, and it should provide a minimum angle of wrap of 120° on the smaller sprocket.

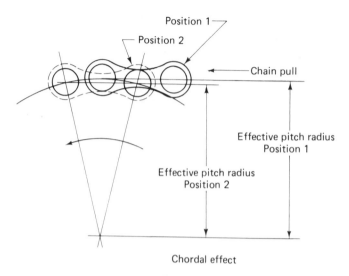

Position 1
Position 2
Chain pull
Effective pitch radius
Position 1
Effective pitch radius
Position 2
Chordal effect

**Figure 7.12**

Roller chain drives have the following advantages: high mechanical efficiency because of low internal friction (of the order of 98% to 99%); no slippage; no initial tensioning requirement; within certain limits any chain length is possible provided it is a multiple of the pitch; and small space required.

They have the following drawbacks: chordal effect, which gives rise to torsional vibrations in the driven shaft; low shock absorption capacity; noise, especially when somewhat worn; and the need for lubrication.

When roller chain wears, it rides progressively higher on the sprocket teeth to accommodate its increased pitch. This, in turn, leads to excessive tension in the chain and causes premature failure. To prevent untimely wear, exposed chain should periodically be removed, cleaned in kerosene, and relubricated.

It is also recommended to "overchain," i.e., to use chain of a somewhat higher rating than necessary. This greatly increases chain life and may in the long run be more economical. The pitch plane of chain mounted on its sprockets should be generally vertical.

## 7.6 SELECTING ROLLER CHAIN

Chain manufacturers' catalogs provide tables that give the rated horsepower of chain of various pitches for different methods of lubrication by entering the revolutions per minute and tooth number of the smaller sprocket. The smallest pitch chain of the required horsepower rating should be selected.

Choose pitches for multiple chain drives on the same basis as that mentioned for multiple V-belts.

**Figure 7.13**
FMC Corporation, Philadelphia, Pennsylvania

(a)

(b)

Odd tooth numbers of sprockets and an even number of pitches in the mating chain are preferred, for this combination will equalize wear on sprocket teeth and rollers.

*Inverted tooth chain* (see Fig. 7.13) was invented in an effort to reduce the considerable noise generated by roller chain (hence its common name *silent chain*).

The hinge point of the links of silent chain is outside the tips of the sprocket teeth which may have either straight or rounded sides. Solid links on both sides of the chain (or a single ridge in the center where space is limited) prevent the chain from leaving the sprockets.

Silent chain is built up from links stamped from sheet metal and can thus be assembled to produce any width up to a certain point. The rated capacity is directly proportional to the width of the chain for a given pitch and revolutions per minute.

Silent chain has reduced chordal effect. This, and the fact that the impact of each link on the sprocket during engagement is much smaller than in roller chain, allows silent chain to run at considerably higher velocities than roller chain. Typical applications include automobile crankshaft drives and silent speed-reducing units.

Selection is mostly similar to that of roller chain.

## 7.7 TOOTHED BELTS OR TIMING BELTS

*Toothed belts or timing belts* are the most recent of all flexible power transmission elements described in this chapter. They combine all the good qualities of both belt and chain and have very few of their drawbacks.

Timing belts are essentially reinforced flat belts with teeth engaging corresponding grooves in the sheaves, which prevents slipping.

Timing belts are usually made from neoprene with synthetic, glass fiber, or steel wire reinforcement and a nylon facing on the inside to resist wear (see Fig. 7.14).

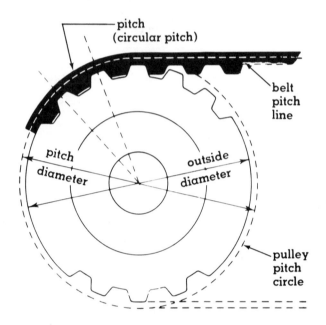

**Figure 7.14**
Uniroyal Industrial Products, Philadelphia, Pennsylvania

A minimum of six teeth must be engaged by the smaller sheave at all times. For this reason, the high rubbing pressures associated with the engagement of gear teeth are absent here. The face of the belt's teeth and the sheave grooves is either flat or curved. Since the pitch plane of the belt lies *outside* the sheave, the face of the teeth cannot have an involute shape (see Chapter 5 on spur gearing).

The sheaves for timing belts do not have crown as those for flat belts, and the belt is prevented from leaving the sheaves by flanges on at least the smaller sheave.

In the original timing belt the shape of the teeth on the belt and those on the sheaves was approximately the same. The load-limiting factor is the maximum permissible shear near the root of the *belt* teeth, which is much lower than its counterpart in the metallic sheave. In an effort to increase the permissible shear load on the belt teeth and thus belt capacity, a more recent design has larger, semicircular teeth on the belt and much smaller mating teeth on the sheaves (see HTD belt in Fig. 7.15).

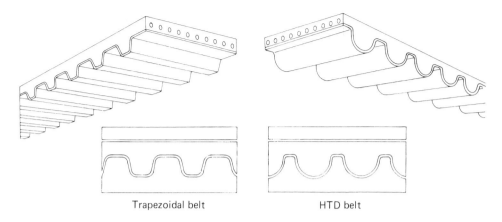

Trapezoidal belt           HTD belt

**(a)**
Machine Design, Cleveland, Ohio

**Figure 7.15(b)**
Uniroyal Industrial Products, Philadelphia, Pennsylvania

Because of their light weight, timing belts may be run at much higher speeds (up to 16,000 ft/min) than V-belts.

Timing belts are available in the following standard pitches: .200, .375, .500, .875, 1.250. See Fig. 7.14 for pitch measurement. Belts for each pitch come in different widths. The width is proportional to the belt's horsepower rating for a given number of revolutions per minute. To select a timing belt for a given design horsepower (see Chapter 4), the correct pitch is found in a chart similar to the V-belt selection chart of Fig. 7.8. Subsequently, a suitable width is selected from a chart showing the available widths for a given pitch. Both charts require the design horsepower and smaller sheave revolutions per minute for input. Lastly, a choice must be made among the available sheaves to obtain the desired speed ratio and among standard belt lengths to obtain the desired center distance.

Timing belts have the following advantages: constant speed ratio (no chordal effect); high mechanical efficiency, similar to that of flat belts; light weight for their capacity; small space required; no lubrication and no other maintenance required; silent operation; negligible stretch; long life. They are thus especially suitable for use in inaccessible places. A small drawback is their low shock absorption capacity, which is still better than that of chain.

Timing belts are at the present time mainly limited by their permissible temperature range ($-65°$ F to $250°$ F). Their rating ranges from fractional to 600 hp.

## PROBLEMS

1. A B60 V-belt (pitch length is 61.8 in.) runs on sheaves having pitch diameters of 7.4 and 12.4, shaft diameters of .625 and .750, and hub diameters of 1.25 and 1.50.

   (*a*) Compute the speed ratio.

   (*b*) Using the equation on page 94, compute the center distance.

   (*c*) Make a half-scale layout of this drive (front view only) as follows. On a size C sheet of vellum draw a horizontal centerline approximately in the center of the paper. Use this line as the line of centers of the drive. Locate the two sheave centers (using the center distance computed before) so that the outsides of the sheaves are approximately the same distance from the sides of the paper. Draw the pitch line of the belt and sheaves as a dash-dot-dash line. Now draw a circle representing the outside of the sheaves as a solid line (see Table 1 in Fig. 7.6 on how to find the OD from the pitch diameter). Now add the two outside tangents representing the outside of the belt. Then draw the insides of the belt (see Table 2 in Fig. 7.6 for belt thickness). Show the hidden portion of the belt and the bottom of the groove (minimum clearance .125 in.) as a dotted line. Now draw a rotated section of the belt in the groove on the outside of each sheave and indicate the correct groove angle (consult Table 2 in Fig. 7.6) and the clearance.

(*d*) Measure the pitch length of the straight portions of the belt and compute the length around the sheaves. Add all these together and compare the result with the given pitch length.

2. What are two drawbacks of flat belts?

3. Explain why a V-belt has a slightly lower mechanical efficiency than a flat belt.

4. What is the one great advantage of internal reinforcement in belts?

5. A pair of sheaves for a C-section V-belt has *outside* diameters of 10.5 in. and 16 in. Find the theoretical speed ratio, i.e., without slip. (*Hint:* Use Table 7.1.)

6. A certain drive, transmitting 12 hp and suitable for either chain or belt, must be exposed to a dusty atmosphere. An exact speed ratio is desirable. What type of drive would you recommend?

7. Two shafts that are to be connected by a belt drive are located in the same horizontal plane. Due to circumstances, the shafts are not entirely parallel. Would you recommend a timing belt in this case? Motivate your answer. If not, what would you recommend?

8. A timing belt pulley has 24 teeth. The angle of wrap (i.e., the arc of contact between the belt and pulley ) is $95°$. Does this angle satisfy the requirement of minimum tooth engagement?

9. In Chapter 4 work was defined as the product of a force and the distance through which it acts and the horsepower as a rate of work performance; thus

$$1 \text{ hp} = \frac{\text{force} \times \text{distance per minute}}{33{,}000}$$

Using this equation, explain why in the curves of Fig. 7.8 it is possible to find four different belt sizes recommended for the transmission of 10 hp. (*Hint:* Distance per minute $= \pi D_p \times \text{rpm}$).

# VARIABLE
# SPEED DRIVES
# AND
# TRANSMISSIONS

*Types, Applications, and Selection. Fluid Power*
*Transmissions.*

Machinery used for many different processes needs a variable input revolutions per minute. The range and number of speeds depend on the nature of the process. For example, in a lathe the optimum spindle speed is a function of the cutting material and the diameter and nature of the workpiece, and in theory at least, an almost infinite number of speeds is required. In other cases, one "high" and one "low" speed may be sufficient (an example is an exhaust fan installation). Most variable input applications have requirements in between these two extremes.

There are two ways in which such speed variation may be effected:

1. By varying the prime mover speed.
2. By varying the speed ratio of the transmission.

Furthermore, the speed of the driven equipment may be varied at constant torque, at constant horsepower, or anywhere between these two conditions. Most loads have constant torque characteristics, which means that, in accordance with the equation we learned in Chapter 4, namely,

$$hp = \frac{\text{torque} \times \text{rpm}}{63,000}$$

when the torque is constant, the horsepower transmitted varies in proportion with the revolutions per minute of the load. Other kinds of loads require constant horsepower within a given speed range, which means that according to the same equation the torque varies inversely with the revolutions per minute of the load. Finally, a few types of driven loads need increases in both torque and horsepower when the revolutions per minute increase.

In most industrial equipment the prime mover is an electric motor of the AC squirrel-cage induction type, which is inexpensive and reliable and needs very little maintenance.

For all practical purposes, however, this is a constant-speed motor, so that when speed regulation of the driven equipment is necessary, a variable speed transmission must be used.

DC motors can be designed for several different methods of electrical speed regulation, each of which has its own characteristic torque and horsepower characteristics that can be matched to the load. Since most power supply systems are three-phase AC, it is necessary to convert this power to DC when such motors are to be used. The total cost of AC-DC conversion equipment, the necessary complicated speed control apparatus and the DC motor itself, is usually much higher than a comparable induction motor that has simple switch gear and a variable speed transmission. In addition, the maintenance costs incurred with DC equipment are higher and the power costs are higher because of the AC-DC conversion losses. For these reasons, DC motors, especially in the larger sizes, are used only when their flexibility of torque, horsepower, and revolutions per minute are of particular advantage, such as in railroad traction motors, elevator and crane drives, and steel mill roll drives, to mention a few. However, modern solid-state DC motor controls are now being combined with a silicon-controlled rectifier (SCR) in one package, which with a matching DC motor that has a base speed adjustable from approximately 4,100 rpm to 2,100 rpm, provides a variable speed drive having capabilities of up to 10 hp [see Fig. 8.1(a)]. Such drives offer several advantages over a mechanical speed reducer driven by a squirrel-cage induction motor which make them competitive in several applications. Some of these built-in advantages are:

1.  Wide speed range [from base speed (see above) down to approximately 40 rpm for the larger sizes].
2.  Control of torque or horsepower or both [see curves in Fig. 8.1(b) and (c)]. This feature also provides flexibility and overload protection for the driven equipment. It also controls acceleration, which, for example, is necessary in a conveyor drive.

(a)

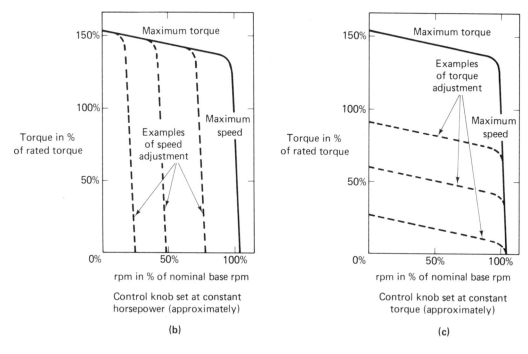

(b)

Torque in %
of rated torque

Examples
of speed
adjustment

Maximum torque

Maximum
speed

rpm in % of nominal base rpm

Control knob set at constant
horsepower (approximately)

(c)

Torque in %
of rated torque

Maximum torque

Examples
of torque
adjustment

Maximum
speed

rpm in % of nominal base rpm

Control knob set at constant
torque (approximately)

**Figure 8.1**
Reliance Electric Co., Inc., Mishawaka, Indiana

3.  Motor overload protection (the motor may be stalled at full torque for indefinite periods of time without damage).

4.  Maintenance of any set speed within 3% of base speed, regardless of load variations, up to maximum load capacity.

5.  Dynamic braking (i.e., self-generated, electrical braking).

6.  Reversibility.

7.  Easy conversion to remote (electrical) control.

The maximum speed of the driver should be the same as that of the load for maximum efficiency. If this is not the case, it is recommended that a V-belt or other drive of suitable speed ratio be installed between unit and load.

Selection of SCR units for *constant torque* applications is done on the basis of maximum horsepower or torque required at the *highest* operating speed of the driven equipment. Selection of SCR units for *constant horsepower* applications is done on the basis of the horsepower or torque required at the *lowest* revolutions per minute of the driven equipment.

For *variable horsepower and torque* applications (both increasing with revolutions per minute), the selection is based on the horsepower or torque requirements at the *highest* revolutions per minute of the driven equipment.

## 8.1  VARIABLE SPEED TRANSMISSIONS

Variable speed transmissions suitable for use with a constant speed motor mostly belong to one of the following main classifications:

1.  Gear drives
2.  Belt and chain drives
3.  Friction drives
4.  Hydraulic torque converters
5.  Fluid power drives

## 8.2  GEAR DRIVES

Gear drives are by nature highly efficient and of the positive (i.e., no slippage) type. Their main drawbacks are that only a limited number of speeds is possible and that change of speed in most types must take place at standstill or at least at zero load. Gear drives are more expensive than belt, chain, or friction drives. They are used in machine tools, automobiles, and other mobile equipment. Types of gear drive are the sliding gear, constant mesh gear, idler gear, and planetary gear. These types are discussed below and are shown schematically in Fig. 8.2.

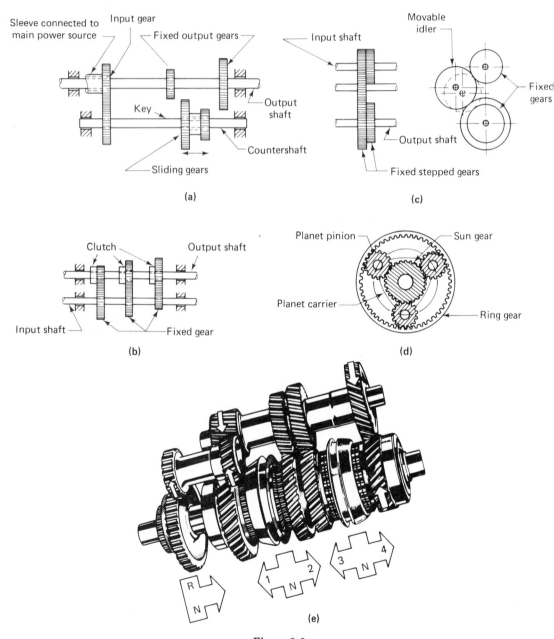

Figure 8.2
Machine Design, Cleveland, Ohio

*SLIDING GEAR.* In the sliding gear type, a number of gears are mounted so that they slide on a countershaft in such a way that they can be selectively engaged with mating output gears mounted on an output shaft. The sliding gear was formerly generally used in automobile manual transmissions, and it is still being used in motorcycles. Changing of gears may be effected while under zero load. To facilitate gear change, the sliding gear was improved upon and evolved into the constant mesh gear.

*CONSTANT MESH GEAR.* As the name implies, the constant mesh gear has a number of gear pairs that are constantly engaged, but the gears on the countershaft are not mounted on splines as in the sliding gear. Each set of gears can be torsionally connected to the countershaft by means of a splined sleeve. The action of the sleeves is such that this connection takes place gradually. Today constant mesh gears are generally used in manual-shift cars.

*IDLER GEAR.* The output shaft carries a stack of gears, fastened to each other and the shaft, of increasing tooth number. The entire stack looks like a stepped cone.

On the input shaft is fastened a gear of a length equal to that of the stack. A lever carrying an idler gear permanently in mesh with the input gear can be moved sideways and inward to permit meshing of the idler gear with any selected gear of the stack. A number of different speed ratios are thus obtained. The idler gear drive is found, for instance, in feed selection boxes of engine lathes. It is suitable for small quantities of power only and should only be engaged at standstill.

*PLANETARY GEAR.* A number of epicyclic gear trains (see Chapter 6) are stacked with the sun gears rotatably mounted on a common shaft, except for the sun gear of the first train, which is fastened to it. The sun gear of the second train is connected to the arms of the first one, and so on. By holding the first, or second, or third, etc., ring gear stationary (for instance, with a band brake) and leaving the other ring gears free to rotate, a number of different speed ratios can be obtained, depending on the individual speed ratios of the component epicyclic trains.

The planetary gear is compact and suitable for the transmission of medium horsepower values. It has the advantage that speed change is smooth and can be effected under load. However, it is expensive and has found only limited industrial application. It is found in certain types of automatic transmissions in passenger cars.

## 8.3 BELT AND CHAIN TYPE DRIVES

Belt and chain type drives all have stepless speed variation, which can take place only when the unit is running.

The group is characterized by the use of *variable pitch sheaves*. Such sheaves are designed so that one or both cheeks can be moved axially and then remain in any selected position. The groove becomes wider or smaller so that the (constant width) belt will ride closer to or further away from the axis of the sheave. When the variable pitch *driver* sheave speed remains constant, this implies that the *belt* speed will either decrease or increase. If belt speed is constant, a variable pitch *driven* sheave will either increase, or decrease in revolutions per minute.

In units having the greatest possible speed range, both sheaves change pitch. One becomes wider and the other becomes narrower. In belt-type drives this is sometimes accomplished by mechanically varying the driver pitch and letting the spring-loaded follower sheave find its corresponding pitch [see Fig. 8.3(a)]. This method is suitable only for relatively small horsepower ratings. In larger belt units, and in the chain type, both sheaves have mechanical pitch control [see Fig. 8.3(b)].

Drives that have one variable pitch and one regular sheave are often used when the reduced speed range is acceptable. Of course, in such units the center distance sheaves must vary in accordance with the pitch change of the driver. This type is inexpensive and adequate for many applications. The pitch change may be positive by means of a screw thread or some other way, or it may be effected merely by moving the driver motor closer to or further away from the driven sheave. In the latter case, the variable-pitch sheave is spring-loaded [Fig. 8.3(c)].

When only one cheek of the sheave is movable axially, the track of the motor must be offset in order to compensate for the sideways motion of the belt on the driver pulley during speed changes [offset visible in Fig. 8.3(c)].

The belts used in all these units are mostly of the wide V-type and have axial grooves on the inside to facilitate bending when the belt rides near the bottom of the adjustable sheave groove.

The belt-type variable speed drive may provide speed ranges of up to 16:1. It is often combined with an electric motor in one package. Efficiency is somewhat less than that of a regular V-belt, but it is still in the 90% range. Slip characteristics are comparable to those of V-belts. Power ratings vary from fractional to approximately 100 hp. Selection for loads having constant torque (i.e., the majority of cases) is on the basis of horsepower transmitted at the highest revolutions per minute of the driven equipment.

In *variable pitch chain* drives the chain links are built up from hardened sheet steel stampings, which are stacked in a direction perpendicular to the direction of motion of the chain. These stampings are permitted some crosswise motion so that they conform to the radial ridges machined in the inside faces of the sheaves [Fig. 8.4(a)]. These ridges are positioned so that any ridge on one face is opposed by a groove on the other face

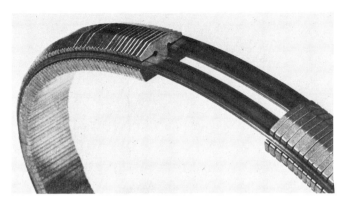

**Figure 8.5**
Van Doorne Transmissie B. V., Eindhoven, The Netherlands

An interesting development in variable-pitch belt drives is the Transmatic System, which was developed and built by Van Doorne's Transmission BV in the Netherlands. In this drive the belt consists of a number of thin, flexible steel belts, each fitting closely inside the next. The belt carries a large number of freely sliding steel links of special shape, as shown in Fig. 8.5. These links transmit the force from driver to driven sheave by pushing instead of pulling, as is the case in conventional drives. This is made possible by the fact that the belt is kept under a tension higher than the pushing force developed by the belt. An integral hydraulic system, pressurized by a gear pump, performs the following functions:

1. Generates inward lateral pressure on one cheek of each sheave, thus tensioning the belt.
2. Controls the desired speed ratio.
3. Controls the starting and overload clutch.
4. Lubricates the belt, the bearings, etc.
5. Removes frictional heat.

A schematic of the hydraulic system is shown in Fig. 8.6, and the complete unit outside of its housing is shown in Fig. 8.7.

Very high efficiencies are obtained with this system, ranging from approximately 87% to 91%, because of the almost frictionless mutual motion of the links. Since the pitch is very small, chordal effect (see under Chain) is almost absent and the noise level is low.

The sheaves are close together, which makes for a compact unit and high permissible speeds.

## 4  FRICTION DRIVES

Under this heading comes a group of mostly metal-to-metal variable-pitch drives. These drives are usually not available in power capabilities over 10 hp, and they should not be used at the same speed setting for any

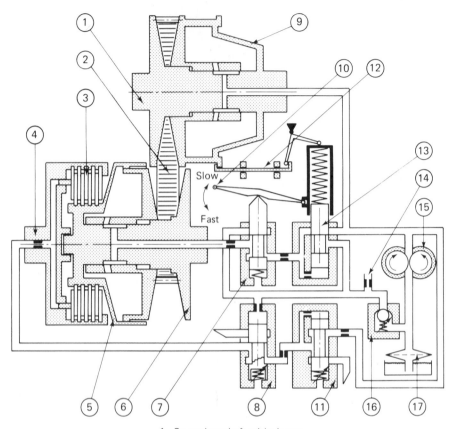

Slow

Fast

1. Secondary shaft with sheave
2. Belt with links
3. Clutch
4. Primary shaft
5. Cylinder of primary shaft
6. Primary shaft with sheave
7. Primary control valve
8. Clutch control valve
9. Piston secondary shaft
10. Control lever
11. Clutch reduction valve
12. Sensor
13. Pressure-limiting valve
14. Lubricating oil line
15. Oil pump
16. Lubricating oil pressure valve
17. Suction filter

**Figure 8.6**
Van Doorne Transmissie B. V., Eindhoven, The Netherlands

length of time, nor should they be subjected to sudden overloads that would cause momentary slipping, since both conditions lead to premature local wear. As with the belt and chain types, speed variation is infinite but possible only during operation. Proper lubrication is essential.

The metal-to-metal action provides compact units with efficiencies of up to 90%.

Several practical designs are available, for example, planetary disc and roller types.

A separate group consists of *viscous drag drives* (Fig. 8.8), in which the contact between the (multiple) discs is provided by an oil film rather

**Figure 8.7**
Van Doorne Transmissie B. V., Eindhoven, The Netherlands

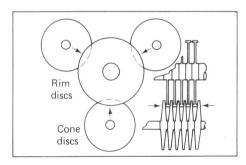

**Figure 8.8**
Machine Design, Cleveland, Ohio

than by metal-to-metal contact. The transmitting force is the shearing force in the oil film at the contact points. While the speed range is fairly small (about 4:1), the drive can not be damaged by overloads nor by long periods of running at the same speed setting, due to the absence of metal-to-metal contact. (In contrast with the first-mentioned type, the discs in the viscous drag drive press onto one another under a very light force only.)

## 8.5  HYDRAULIC TORQUE CONVERTERS

A hydraulic torque converter is essentially a hydraulic coupling (see Chapter 12) with the addition of a stator in the flowpath between the turbine and the impeller [see Fig. 8.9(a)]. In the sketches of the flowpaths

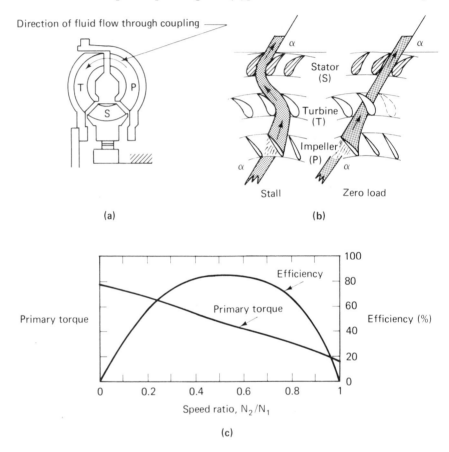

(a)

(b)

(c)

**Figure 8.9**
Machine Design, Cleveland, Ohio

at stall and at zero load [Fig. 8.9(b)] note that the stator provides a constant angle $\alpha$ (alpha) of fluid entry into the impeller blades. This feature provides high efficiency over a fairly wide range [see the curve in Fig. 8.9(c)].

The hydraulic torque converter provides the same advantages as the hydraulic coupling but with the added feature of variable speed. For a given, constant revolutions per minute input, the output torque varies inversely with the revolutions per minute in the approximately horizontal portion of the efficiency curve, say between $N_2/N_1$ = .3 to .75.

## 8.6  FLUID POWER DRIVES

In a fluid power drive a positive displacement pump (i.e., with pistons) moves a fluid under high pressure through a positive displacement fluid motor, thus transmitting energy. In the design shown schematically in Fig. 8.10 the pump and motor are both of the swashplate type. The

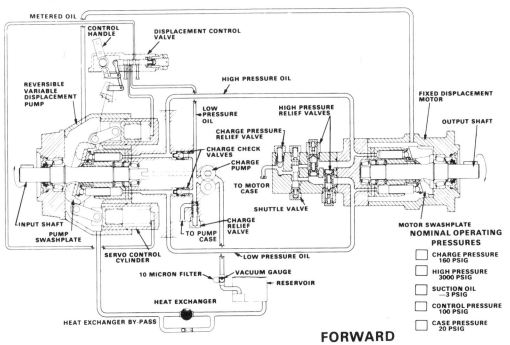

### FALK FLUID POWER DRIVES

TYPICAL HEAVY DUTY VARIABLE PUMP–FIXED MOTOR TRANSMISSION SCHEMATIC

**FORWARD**

Figure 8.10
Falk Corporation, Milwaukee, Wisconsin

swashplate of the pump is hinged on trunnions in such a manner that its angle with the normal plane of the pump shaft can be varied from $0°$ to $18°$ on each side of the normal plane (Fig. 8.11). The motor is similar

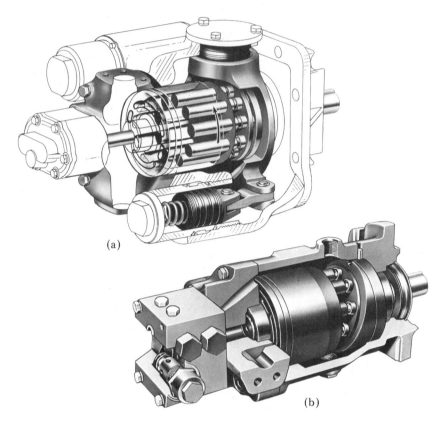

(a)

(b)

**Figure 8.11**
Falk Corporation, Milwaukee, Wisconsin

to the pump except that the angle of the swashplate is fixed. The motor thus requires a fixed amount of fluid to make one revolution. By reducing the angle of the pump swashplate, the *quantity* of fluid passing through the motor during a given length of time is reduced, and the motor revolutions per minute are thus reduced in proportion. By moving the pump swashplate to an angle on the other side of the normal plane of the pump shaft, the direction of fluid flow is reversed: the port that was first the inlet port of the pump now is the exhaust port, and vice versa. The same reversal takes place in the motor.

Not only is the driven equipment quickly reversed in this manner, but dynamic braking is also provided. Since the motor has positive displacement, it cannot rotate when the flow of fluid is stopped, as is the case when the pump swashplate is placed at an angle of 0° with the normal plane of the pump shaft.

A reservoir is placed beneath the pump unit and is of sufficient capacity to permit the separation of possible air bubbles and settling of any dirt. A charge pump keeps the entire system filled with oil. The heat from friction losses (both hydraulic and mechanical) is removed by making the oil flow through a heat exchanger. In the unit shown the maximum continuous design system pressure is 3,000 psi. Relief valves available in 500 pound increments up to 5,000 psi provide overload protection. As will be seen from the typical horsepower vs revolutions per minute curve in Fig. 8.12, the unit has an output of 32 hp. At a typical input of 40 hp, it has an efficiency of 32/40 × 100% = 80%.

Design load combinations of horsepower and revolutions per minute should fall within the white region in Fig. 8.12. It will also be seen in the curve that when revolutions per minute are reduced, by increasing torque from the maximum value of 1,684, the maximum horsepower transmitted is constant, with the system pressure rising from 1,916 psi until the maximum permitted value of 3,000 psi is reached. When revolutions per

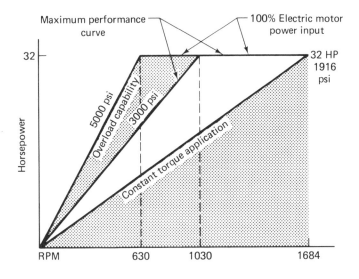

**Figure 8.12**
Falk Corporation, Milwaukee, Wisconsin

minute (and thus fluid flow) are further reduced, the maximum horsepower transmitted will be reduced in proportion with the revolutions per minute. This is in accordance with the equation

$$\text{hp} = \frac{\text{force} \times \text{distance}}{\text{time unit}}$$

where force is the constant pressure of 3,000 psi and distance is proportional to the quantity of hydraulic fluid circulated per time unit (curve marked "maximum performance"). At constant torque applications, the delivered horsepower varies directly with the revolutions per minute, as is seen from

$$hp = \frac{torque \times rpm}{63,000}$$

or

$$\frac{constant\ torque}{63,000} = another\ constant = \frac{hp}{rpm}$$

For constant torque applications, selection of a fluid power drive unit is therefore based on the *output torque at maximum speed* (i.e., at maximum horsepower). In constant horsepower applications, however, the unit is selected on the basis of *required output horsepower at the lowest speed of the range* (i.e., at maximum torque).

Fluid power systems have the unique advantage of mounting flexibility, in that no alignment of any kind is necessary between driver unit and load. With flexible hosing between the pump and motor, limited motion between the two in all directions is possible.

Apart from the advantages already noted, such as instant reversibility and dynamic braking, fluid power systems have a low weight/horsepower ratio. This feature is of particular advantage in mobile equipment.

In a separate class is the unusual and interesting ZeroMax drive (Fig. 8.13), which consists essentially of three parallel five-bar linkages on common input and output shafts, with their driving eccentrics at 120°, 240°, and 360°. The length of two intermediate links is variable and permits output variations from 0 rpm to 400 rpm with 1,800 rpm input. An overrunning clutch (see Chapter 12) at the driven end of each linkage provides rotation in one sense (direction) only.

The drive is silent and compact and it has an integral mechanical overload protection. The zero output feature permits gradual starts. However, it is suitable for low horsepower values only (up to 1.5 hp), and the output revolutions per minute are subject to minor ripples which become noticeable at low input revolutions per minute values.

## PROBLEMS

1. In which two ways may the speed variation of a driven load be effected? Give examples of each.
2. Why is the most common prime mover in industry the squirrel-cage induction motor?
3. What are the advantages of SCR drives over variable-pitch belt drives having squirrel-cage induction motors?

4. Why is stepless speed variation impossible with gear-type drives?

5. A variable-pitch drive has a speed ratio range of 1 to 14. Would you recommend using it with a load that requires speeds that vary from 750 rpm to 900 rpm?

7. What are the differences in performance between variable-pitch belt drives and variable-pitch chain drives?

8. What are the advantages and disadvantages of fluid power drives?

(a)

(b)

**Figure 8.13**
ZeroMax Industries, Inc., Minneapolis, Minnesota

# 9

# BEARINGS

*Radial, Axial, and Linear Bearings. Journal Bearings and Rolling Contact Bearings. Selecting Bearings from Commercial Catalogs.*

A bearing is a support system for rotating, oscillating, or translating machine elements in which friction has been reduced to a minimum.

*Plain bearings* provide *sliding* support and friction is reduced by a lubricant, which may be a fluid, a semifluid, or a solid.

In *bearings with rolling contact* the surfaces in relative motion are separated by balls, rollers, or needles (i.e., rollers which are very long with respect to their diameter).

## 9.1 PLAIN BEARINGS

1. Journal bearings (sleeve bearings) support a rotating or oscillating shaft.
2. Thrust bearings support an axial load on a rotating or oscillating shaft.
3. Line or guide bearings guide and support translating motion.

All these bearings must be lubricated. The lubricant is mostly oil (in special cases water), sometimes grease, and (rarely) a compressed gas such as air. Certain bearing materials provide self-lubrication.

Ideally, a layer of lubricant separates moving surfaces. In that case, the only resistance to motion is that caused by the shear forces between adjacent lubricant layers.

*JOURNAL BEARINGS.* According to the *hydrodynamic theory of lubrication*, a plain, fluid-lubricated journal bearing may be in one of the three distinct states shown in Fig. 9.1.

In Fig. 9.1(a) the shaft is at rest and is supported by the inside bottom of the bearing in metal-to-metal contact. Oil lubricant fills the space between the shaft and bearing. In the sketches the difference between shaft *OD* and bearing *ID* (the "diametral clearance"), has been grossly exaggerated for clarity.

At the beginning of the shaft rotation in Fig. 9.1(b) the shaft first rolls up the inside wall of the bearing but at one point it slides back by gravity. When that happens, some oil separates the shaft from the bearing and lubrication begins.

In Fig. 9.1(c) the shaft has reached such a speed that oil adhering to its surface actually lifts the shaft up above the bearing surface by a wedging action, so that metal-to-metal contact no longer exists.

The slant of the axis in Fig. 9.1(c) represents the direction of the resultant of the gravity force and the sideways force exerted by the oil entering-wedge *w*, on the shaft. The wedged-in oil partly escapes on either side of the bearing and partly passes through the narrowest section of the clearance ring. The oil pressure distribution around the shaft is suggested by the dotted line *L*. The longitudinal pressure distribution at the point of highest pressure is indicated in Fig. 9.2. This description is valid only when there are no reaction forces on the shaft caused by driven equipment. When such exist, the *resultant* of gravity, reaction, and wedging forces determines the position of the axis in Fig. 9.1(c), and thus the high and low pressure distribution around the bearing. This fact should be kept in mind when the location of oil supply holes or grooves is decided upon, since they should always be positioned at the point of lowest hydrodynamic

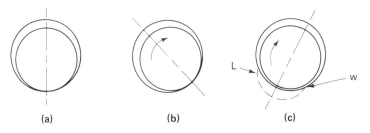

(a)          (b)          (c)

**Figure 9.1**

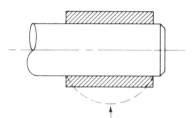

Typical longitudinal hydrodynamic
oil pressure distribution

**Figure 9.2**

pressure. In small equipment and installations the reaction force is usually much larger than the gravity force.

The condition described above, called *complete or fluid lubrication*, is often not met in practice. In many instances, *partial or boundary lubrication* takes place where hydrodynamic lubrication cannot occur, for example, in oscillating shafts and in guide bearings for reciprocating motion.

Oil may be automatically supplied to the bearing by partial submersion of the shaft, by wick, or by chain or ring from a sump below the shaft (see Fig. 9.3). Other methods exist.

In cases of heavy bearing loads, it may be necessary to provide lubricant under pressure. Lubricant pressure $P = \dfrac{W}{LXD}$ in which $W$ = total load on bearing, $L$ = length and $D$ = inside diameter of bearing.

The shearing action of lubricant layers in hydrodynamic lubrication produces frictional heat, which is dissipated to the surroundings, mainly by radiation but to a lesser degree also by conduction and convention.

The quantity of heat generated in journal bearings is proportional to the shaft's revolutions per minute and to the *cube* of the bearing's diameter. This explains why large-diameter, high-speed bearings require external cooling of the bearing or the lubricant or both, in combination with a pressurized lubrication system.

Generally speaking, when the oil temperature in a bearing stabilizes at 140°F to 160°F or higher, oil cooling becomes mandatory, since the petroleum-based oils commonly used for lubrication oxidize rapidly above that range. Also, the viscosity (thickness) of the oil decreases with increasing temperature, which reduces the load capacity of the film.

The computation of *journal bearing load capacity* is fairly complicated and beyond the scope of this text. It depends on the nature and surface condition of shaft and bearing materials used and is a function of bearing length, diameter, revolutions per minute, oil viscosity, and diametral clearance (bearing diameter minus shaft diameter). Table 9.1 shows typical load ranges encountered in various journal bearing applications. In general, manufacturers' recommendations of load rating should be followed.

Shafts should be considerably harder than the bearings and should preferably be case-hardened (surface-hardened).

TABLE 9.1

| Type of bearing | *Permissible pressure, psi, of projected area | Type of bearing | *Permissible pressure, psi, of projected area |
| --- | --- | --- | --- |
| Diesel engines, main bearings . | 800-1,500 | Steam turbines and reduction | |
| Crankpin . . . . . . . . . . . . | 1,000-2,000 | gears . . . . . . . . . . . . | 100- 220 |
| Wrist pin . . . . . . . . . . . . | 1,800-2,000 | Automotive gasoline engines, | |
| Electric motor bearings . . . . . | 100- 200 | main bearings . . . . . . . . . . | 500- 600 |
| Marine Diesel engines, main | | Crankpin . . . . . . . . . . . . . | 1,500-2,000 |
| bearings . . . . . . . . . . | 400- 600 | Air compressors, main bearings . . | 120- 240 |
| Crankpin . . . . . . . . . . . | 1,000-1,400 | Crankpin . . . . . . . . . . . . | 240- 400 |
| Marine line-shaft bearings . . . . | 25- 35 | Crosshead pin . . . . . . . . . | 400- 800 |
| Steam engines, main bearings . | 150- 500 | Aircraft engine crankpin . . . . . | 700-2,000 |
| Crankpin . . . . . . . . . . . . | 800-1,500 | Centrifugal pumps . . . . . . . . . | 80- 100 |
| Crosshead pin . . . . . . . . | 1,000-1,800 | Generators, low or medium speed | 90- 140 |
| Flywheel bearings . . . . . . | 200- 250 | Roll-neck bearings . . . . . . . . | 1,500-2,000 |
| Marine steam engine, main | | Locomotive crankpins . . . . . . . | 1,500-1,900 |
| bearings . . . . . . . . . . | 275- 500 | Railway-car axle bearings . . . . . | 300- 350 |
| Crankpin . . . . . . . . . . . | 400- 600 | Miscellaneous ordinary bearings . . | 80- 150 |

*The permissible pressure is the *bearing pressure*, which is equal to the oil pressure p in the bearing (See p. 130).

Reprinted with permission from *Mark's Handbook for Mechanical Engineers*, McGraw Hill.

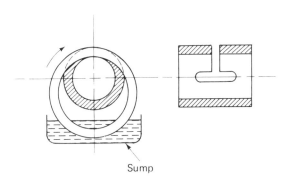

Sump

**Figure 9.3**

In general, the smoother the mating surfaces of shaft and bearing, the greater the load-bearing capacity and the lower the required oil viscosity (thickness). Low viscosity also implies low friction losses. This general rule is valid for hydrodynamic lubrication only. In cases of boundary lubrication, some microscopic surface roughness actually helps to retain the lubricant.

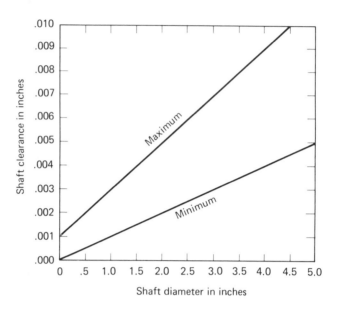

**Figure 9.4**
Boston Gear, Quincy, Mass

The amount of diametral clearance needed depends mainly on load, revolutions per minute, diameter, and oil viscosity. The curves in Fig. 9.4 give the recommended maximum and minimum clearance values vs shaft diameter for average conditions.

Recommended length/internal diameter ratios ($L/D$) for journal bearings lie between .8 and 1.3, but smaller ratios are often unavoidable because of space limitations. It should be noted that, all other things remaining equal, decreasing the $L/D$ ratio will reduce the load capacity of journal bearings by a factor greater than the ratio reduction. This is a result of the increasing influence of oil leakage from the ends of the bearing. See Chapter 10 for other data on sleeve bearings.

*Journal bearing materials* should preferably have the following characteristics: good compressive and fatigue strength; good heat conductivity; chemical inertness to shaft and housing material as well as lubricants; and ductility to conform to the shape of housing and shaft and to permit embedment of foreign particles that would otherwise scour the shaft.

The most commonly used, general-purpose bearing materials come from a group of *bronze alloys* (mainly consisting of copper and tin) that conform well to the above requirements. Other bearing materials are babbitt, silver, cast-iron, sintered materials, and nonmetallic compounds. These bearing materials are discussed below.

*Babbitts* (alloys consisting mainly either of tin or lead and having poor fatigue characteristics) and *silver* are mostly used in thin layers with cast-iron or steel backings. Silver is used for heavy-duty applications.

*Cast-iron* bearings are low-cost and operate well with hardened steel shafts, but they have low embeddability and poor heat conductivity.

*Rubber* bearings are used for underwater applicatons, such as in hydraulic turbines, pumps, and ship propeller shafts. The water serves both as lubricant and as coolant. Shafts must be made from non-corroding materials.

*Sintered bearings* are made by compressing metal powders in a die and subsequently sintering the compacts in an oven. (Sintering involves heating the constituents to a temperature somewhat below the melting point. This causes the powder particles to weld together.) The powders used are mostly copper and tin mixtures in the same ratios as used in bronzes, and sometimes iron. In both cases, other powders are sometimes added to impart some desirable property.

Sintered bearings are more or less porous (depending on the pressure used in their manufacture), and they may be charged with oil. They thus become self-lubricating and suitable for use in places where regular maintenance is difficult or impossible. They have found widespread use in small electric motors and appliances as well as in bearing units for shafting and in other, larger equipment (see Chapter 10). They are often made with a spherical exterior so as to permit self-alignment.

Because of their porosity, the load capacity of sintered bearings is below that of the solid bronze type. The chart in Fig. 9.5 gives allowable pressure in pounds per square inch of *projected* bearing area (length times diameter) vs shaft revolutions per minute for various shaft diameters.

Sleeve bearings in general are usually held in place by a force (interference) fit. Since bearings close a little when they are pressed into place, the manufacturers' hole size recommendations should be followed to the letter to avoid the necessity of having to size the bearing ID later.

Other self-lubricating bearings are made from *nonmetallic compounds* such as teflon, nylon filled with graphite or molybdenum disulphide, or similar materials. In these bearings lubrication is effected by *transfer* of some of the bearing material to the shaft. The low coefficient of friction between the transferred layer and the parent bearing material accounts for the low friction values of these bearings. They are used in textile machinery, food-processing equipment, business machines, and several other applications in which the presence of oil or grease is objectionable or undesirable.

Such bearings are often selected on the basis of maximum permissible values of $PV$, in which $P$ = load in lb/in.$^2$ of projected bearing area (see sintered bearings) and $V$ = shaft surface velocity in feet per minute. $V$ is

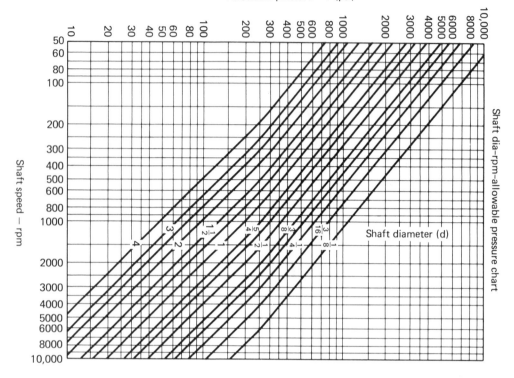

Allowable pressure — P (psi)

Shaft speed — rpm

Shaft dia-rpm-allowable pressure chart

Shaft diameter (d)

**Figure 9.5**
Boston Gear, Quincy, Mass

numerically equal to ($\pi D \times$ rpm/12). It will be clear that the lower the shaft speed, the higher the permissible bearing load, and vice versa.

Maximum permissible $PV$ values for a typical material, consisting mainly of teflon, are as follows:

1.  20,000 for continuous, unlubricated service.
2.  50,000 for intermittent service *or* continuous service while lubricated or submerged in a fluid.

*THRUST BEARINGS.* Thrust bearings for relatively low loads and speeds may consist of a shoulder on a shaft that rotates under axial pressure against a flanged sleeve bearing. Higher loads may be accommodated in a similar manner by *multiple shoulders* that rotate against *horseshoe-shaped bearing pads.*

In the efficient *Kingsbury thrust bearing*, tiltable shoes are located in a ring around the shaft and bear upon a single shoulder (see Fig. 9.6). When the shaft is rotated and presses against the pads, a wedge of oil builds up between the tilted shoes and the shoulder in accordance with hydro-

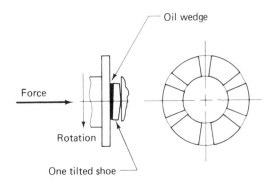

Oil wedge

Force

Rotation

One tilted shoe

**Figure 9.6**

dynamic theory. This reduces the friction to that of the shear between the adjacent oil layers in the wedges. An adequate flow of oil on the active face of the shoulder is maintained so that the wedges will not collapse.

*LINEAR OR GUIDE BEARINGS.* Linear or guide bearings are often mated with an inverted V-shape to reduce the effects of wear. An example of such a construction is found in the ways of quality lathes and in crosshead guides in reciprocating, double-acting air compressors.

Since most guide bearings are used for reciprocating motion, which prevents the formation of a separating oil wedge, they are often scored or finely grooved to aid the retention of lubricant, which is usually grease. In these bearings, *boundary* or *semifluid* lubrication takes place.

All plain, rotary bearings that do not receive an externally imposed oil pressure prior to startup experience some wear during each start before the hydrodynamic oil pressure in the bearing builds up and lubrication begins. For this same reason, considerable friction may be present at the beginning of rotation. This is known as the *slipstick effect.*

For applications having many starts and stops, when starting friction must be low and where oil lubrication is difficult, bearings with rolling contact may be called for. These bearings are discussed in the next section.

## 9.2  BEARINGS WITH ROLLING CONTACT

In bearings with rolling contact the shaft is directly or indirectly supported by rolling elements such as balls, rollers, or needles, which reduce friction to that caused by elastic deformation of the surfaces in rolling contact, sliding friction of rolling elements with cages, or retaining rings, or with one another, and some shear of lubricant.

With respect to the bearing bore (i.e., the shaft diameter), the coefficient of friction for all types is less than .005; however, the use of lubricant-retaining, or dirt-excluding, seals may increase the friction considerably and bring it above that of comparable (operating) hydrodynamic bearings.

The starting friction is considerably lower than that of sleeve bearings and, to all practical purposes, the friction is not increased by load.

*Ball bearings* have rolling elements in the form of balls, which in all but the most inexpensive types are retained in cages, separators or retainers, and inner and outer races.

*Roller bearings* have mainly cylindrical or conical rollers instead of balls, but otherwise they are similar to ball bearings.

*Needle bearings* usually have neither an inner race nor a cage. The needles are retained by integral flanges on the outer race.

*BALL BEARINGS.* The inner and outer races of ball bearings have a rounded, annular groove to provide a track for the balls. The sectional radius of curvature is a little larger than the radius of the balls (see Fig. 9.7(a)).

Cages or separators are mostly made from steel stampings, but bronze, reinforced phenolic, or other desirable materials are also used, depending on the service for which the bearing is intended.

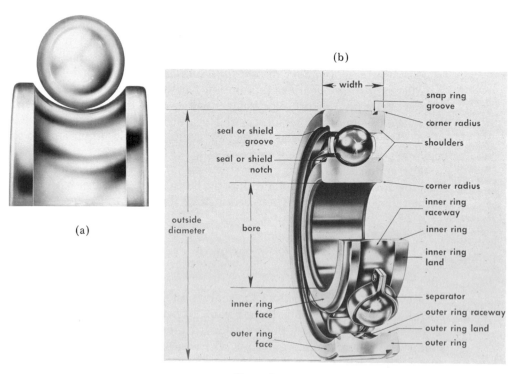

Figure 9.7
New Departure/Hyatt Bearings, Inc., Sandusky, Ohio

*Radial bearings* are designed for mainly radial loads, although the deep-groove types also have considerable axial capacity [see Fig. 9.7(b)].

*Angular contact bearings* are designed such that the centerline of contact between the balls and races is at an angle with the plane of the bearing (see Fig. 9.8). This type is suitable for unidirectional axial (thrust) as well as radial loads. It is often used in conjunction with another, similar bearing on the same shaft in such a manner that the contact centerlines are opposed. This assembly permits preloading which prevents axial and radial play in the shaft.

*Thrust bearings* have two identical, ring-shaped flat races with a concentric groove in one face. The grooves face one another and contain the balls (see Fig. 9.9). Thrust bearings will not support radial loads.

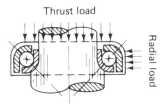

**Figure 9.8**
New Departure/Hyatt Bearings, Inc., Sandusky, Ohio

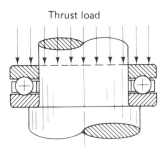

**Figure 9.9**

*LOAD CAPACITY OF BEARINGS WITH ROLLING CONTACT.*
In these bearings the rolling elements (balls, rollers, and needles) in theory make *point or line contact* with the races (or race and shaft in the case of needle bearings).

In practice, because of the elastic deformation at the contact points, there exists a small *contact area* of approximately elliptical shape (see Fig. 9.10). Nevertheless, since $S = F/A$ (stress equals the force divided by the area), and since $A$ is very small, extremely high stresses (of the order of 300,000 psi to 500,000 psi are generated in the contact areas even at moderate loads. These stresses travel around the races and balls during

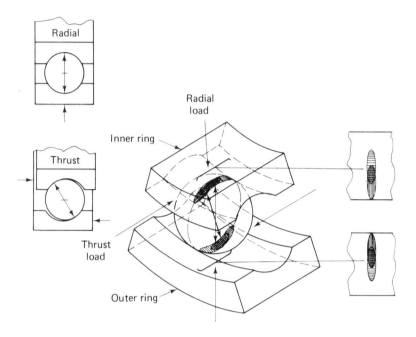

**Figure 9.10**
New Departure/Hyatt Bearings, Inc., Sandusky, Ohio

rotation of the bearing, and thus they vary from zero to a maximum and back to zero at any point of the load contact area.

Such cyclical stresses cause fatigue failures when they are above the endurance limit of the material, which is the case in all bearings with rolling contact operating at their rated loads. For this reason, they all have a finite life, in contrast with hydrodynamic bearings which, if kept running and properly lubricated, last indefinitely because there is no metal-to-metal contact. The life of rolling contact bearings is not the same for each individual bearing from a group but varies statistically. It is customary for manufacturers to give load ratings in their catalogs such that 90% of the group will have a minimum life of 1 million revolutions at that rated load.

Service factors should be applied to the basic load values, as discussed in Chapter 4, before bearings are selected from catalogs.

It is also possible to express bearing life as a number of hours of operation at a certain number of revolutions per minute, the product of hours, revolutions per minute and 60 being one million (for example, $8\frac{1}{3}$ h at 2,000 rpm, etc.). From this it is obvious that at a given load the revolutions per minute are inversely proportional to the life; in other words, doubling the revolutions per minute means halving the life, and so on.

Changes in load, however, have far greater influence on bearing life, since for ball bearings, the *life varies inversely as the third power of the load* (load$^3$). For roller bearings, the power is 3.3 (load$^{3.3}$). For ball bearings, this relationship may be expressed in the following equation:

$$\frac{(\text{rated load})^3}{(\text{actual load})^3} = \frac{\text{actual life, revolutions}}{10^6 \text{ (i.e., rated life, revolutions)}}$$

The following example will clarify the use of this equation.

Given: A radial ball bearing has a rated load capacity of 130 lb for a life of $10^6$ revolutions.

Required: The life if the load is increased from 130 lb to 150 lb and all other factors remain the same.

Solution:

$$\frac{(130)^3}{(150)^3} = \frac{\text{actual life, revolutions}}{10^6}$$

Actual life $\approx 650{,}000$ revolutions

The static (nonrotating) load capacity of bearings with rolling contact is generally lower than the dynamic (rotating) capacity due to the possibility of permanent indentation of the races ("Brinelling") and/or flattening of the rolling elements. For applications in which the static loads are high, it is wise to consult with the manufacturer as to what type of bearing and what rating should be used.

As mentioned before, radial ball bearings have some axial (thrust) capacity in addition to their radial capacity (see Fig. 9.10). Manufacturers' catalogs provide simple computational methods to convert any axial load into an *equivalent radial load*. If the load is axial only, the equivalent radial load can be directly used as a basis for selection. If an axial load coexists with a radial load, the *equivalent* radial load is added to the *actual* radial load. The sum is then used for selection.

The *materials* used in manufacturing rolling contact bearings are mostly vacuum-degassed, high-purity steels containing approximately 1% each of carbon and chromium. Sometimes nickel and molybdenum are added. Stainless steels are used in bearings that must operate under corrosive and/or high-temperature conditions.

The materials should be capable of being heat-treated to a very high hardness and at the same time they should have a high endurance strength (i.e., resistance to fatigue stresses), two properties which often counteract one another. Failure of rolling elements and races is typically indicated by flaking of particles off the surface ("spalling"). The flaking is caused by fatigue of the surface layers.

Standard boundary dimension groups of ball bearings are shown in Fig. 9.11.

*Lubrication* of bearings with rolling contact is mainly by grease. Seals are often used to keep dust or grit from entering the bearing and also to retain lubricant. (See Fig. 9.11).

When space permits, bearings may be run partially submerged in oil. This provides excellent lubrication and is particularly well-suited to high revolutions per minute conditions, but the oil level must be watched carefully because too high a level may lead to excessive churning which generates heat, and too low a level may deprive the bearing of lubrication.

Forced oil circulation may be necessary in large, high-speed installations to remove the heat generated in the bearing.

Ball and roller bearings are most often force-fitted onto shafting and slide-fitted into housings or mountings, thus permitting the outer race to float.

Manufacturers' catalogs usually provide recommended shaft and hole dimensions and tolerances. The shaft dimensions are especially important because they affect the critical clearances between races and rolling elements, due to the expansion of the inner race when pressed onto the shaft. Too much expansion of the inner race will cause early bearing failure due to excessive compressive stresses within the races and the captive rolling elements.

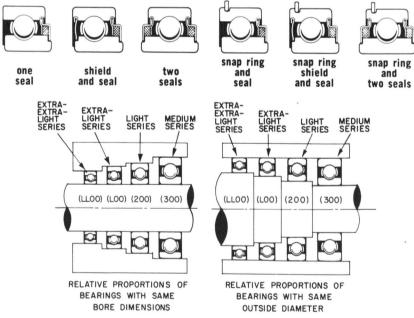

**Figure 9.11**
New Departure/Hyatt Bearings, Inc., Sandusky, Ohio

*ROLLER BEARINGS.* Since roller bearings have line instead of point contact, they have much greater load capacity than ball bearings of similar exterior dimensions. Bearings that have *cylindrical rollers* have no axial load capacity, but bearings with *conical rollers* have axial as well as radial load capacity. The latter are used in the same sort of applications that angular ball bearings are used, that is, where zero play is required, for example in surface grinder spindles.

*Self-aligning* roller bearings have a spherical outer race and rollers and inner race to match. They also take some axial load. There are many other types of roller bearings designed for specific applications on the market. Figure 9.12 shows only those discussed here.

Most of the information given for ball bearings also applies to roller bearings. Therefore, selection is made on the same basis.


*NEEDLE BEARINGS.* Of all bearings with rolling contact, needle bearings provide the greatest load capacity in the smallest space (see Fig. 9.13). They have no cage and no separator, and the needles rub against one another, which, of course, increases friction and makes this type less suitable for higher revolutions per minute applications.

It is almost mandatory that the portion of the shaft that serves as the inner race be case-hardened, since very short bearing life would be obtained otherwise.

The outer race of needle bearings is generally very thin. It has no stable cylindrical configuration of its own, and it assumes the shape of the hole into which it is pressed. The roundness of that hole is therefore critical.

Some newer designs of needle bearings have either a retainer or an inner race, or both; however, the latter type is technically a roller bearing with needle-type rollers.

Needle bearings are mostly used in high-load, low revolutions per minute applications in which space is at a premium. Lubrication is almost always by grease.


*LINEAR MOTION BEARINGS WITH ROLLING CONTACT. Linear ball bearings* (Fig. 9.14) are used with circular-section shafts. Balls are located in oblong circuits, one side of which is exposed to the shaft.

Linear ball bearings combine low friction characteristics with high load capacity and small space requirements. Zero clearance fits are available when this feature is essential, for example, in diesets. As with needle bearings, the shaft should be case-hardened.

*Roundway bearings and ways* (Fig. 9.15) provide greater load capacity than do ball bushings and may offer savings over flat ways when operated with inexpensive centerless ground and hardened bars.

Sleeve bearings and bearings with rolling contact have different characteristics and properties. A comparison of these is given in Table 9.2.

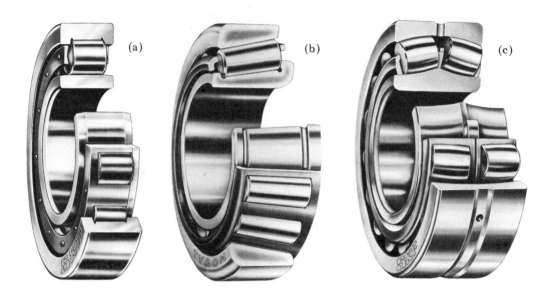

Figure 9.12
New Departure/Hyatt Bearings, Inc., Sandusky, Ohio

Figure 9.13
The Torrington Company, Torrington, Connecticut

Figure 9.14
Thomson Industries, Inc., Manhasset, New York

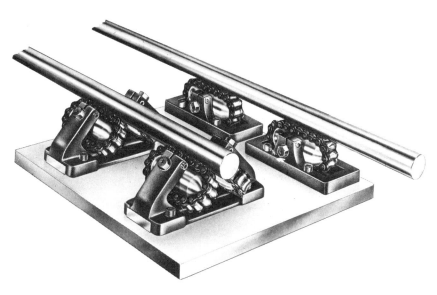

**Figure 9.15**
Thomson Industries, Inc., Manhasset, New York

## TABLE 9.2

*Comparison of properties of journal bearings
and bearings with rolling contact*

| Metal journal bearings | Bearings with rolling contact |
|---|---|
| Relatively high starting friction. Running friction mostly somewhat higher than rolling contact bearings. | Little difference between starting and running friction. |
| Except for sintered bearings, need some form of external lubricant supply. | May have lubricant sealed in for life. |
| Magnitude of the load influences the needed oil viscosity which in turn influences friction. | Needed lubricant viscosity not affected by load. Little variation of friction with load changes. |
| Unlimited life if kept running and properly lubricated. | Finite life influenced by load and revolutions per minute. |
| Generally quiet. | May be noisy. |
| Need little radial space. | Ball and roller bearings need considerable radial space. |
| Inexpensive. | Expensive. |
| Can often be repaired. | Cannot be repaired. |

## PROBLEMS

1. Fig. 9.16 shows a typical engineering department layout (at a reduced scale) for a polishing disc drive in which some modificatons have been made, as indicated by dimensions marked NTS (not to scale). Make a new layout of this drive and show everything at the correct scale. Use size B vellum. Find the belt groove dimensions in the tables in Chapter 7 and draw a pulley to scale. Start with the pitch diameter and then work outward and inward. Do not dimension the drawing but make assembly notes as needed.

2. A gear-type speed reducer is to transmit 1.5 hp with a service factor of 1.2. Driver rpm = 1,200 and driven rpm = ±500. The center distance is to be between 5 and 6. The following 16-diameter pitch, $20°$ -pressure angle gears have been selected as suitable:

    (a) Driver: .75 face width, 48 teeth. Steel, plain type. Hub width .63, hub diameter 1.50.

    (b) Driven: .75 face width, 112 teeth. Cast-iron, spoke type. Hub width .75, hub diameter 1.75.

    (c) Driver shaft: .7500 nominal diameter; ends .5000 nominal diameter.

    (d) Driven shaft: .7500 nominal diameter; ends .6250 nominal diameter.

    The bearings in Fig. 9.17 have been selected as suitable for load and revolutions per minute. Fig. 9.18 gives the general arrangement of the parts in the lower part of the housing. Make a layout at full scale on size C vellum. Start with the gears and then draw the lower housing around them. Show the gear pitch as dotted lines and the gear outline as a solid line. Sketch a few teeth as indicated. The layout must show the top view, elevation, and side view. Make the housing wall thickness .25, the flange width 1.25, the corner radius inside .5, and the outside concentric with the inside. The minimum clearance around the gears is .25. The draft of the cast iron housing is 1° all around. Compute center distance. Do not dimension drawing. Scale or estimate dimensions not given or implied.

3. A certain bearing costs $40 and lasts 500 h. Its installation costs $14. A salesperson informs you that a similar bearing from his company costs $45 but lasts 650 h. To install the bearing requires reboring of the housing. The total installation costs $30. Which would you select? (*Hint:* Add the installation costs to the price of the bearing and divide the total by the number of hours of life.)

4. A journal bearing has a length of 5.75 in. and a shaft (i.e., journal) diameter of 4.5 in. When the oil pressure is 120 psi, what is the total load on this bearing?

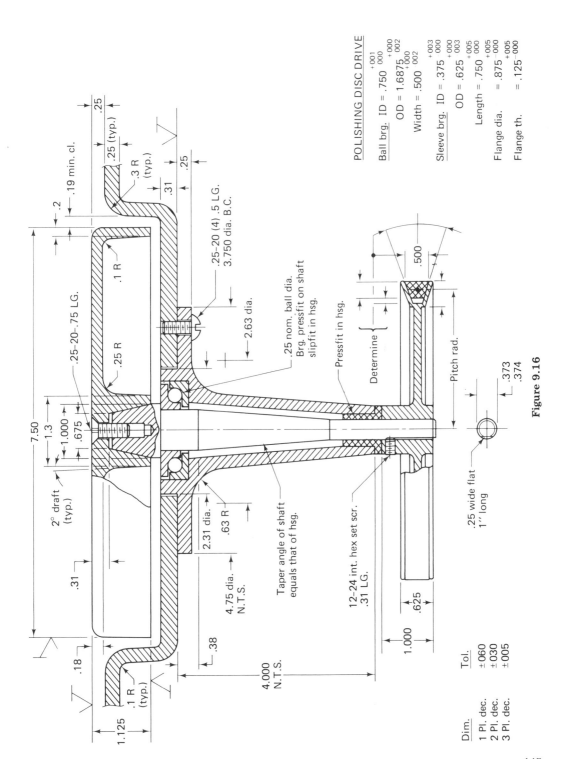

**Figure 9.16**

POLISHING DISC DRIVE

Ball brg. ID = .750 $^{+.001}_{-.000}$

OD = 1.6875 $^{+.000}_{-.002}$

Width = .500 $^{+.000}_{-.002}$

Sleeve brg. ID = .375 $^{+.003}_{-.000}$

OD = .625 $^{+.000}_{-.003}$

Length = .750 $^{+.005}_{-.000}$

Flange dia. = .875 $^{+.005}_{-.000}$

Flange th. = .125 $^{+.005}_{-.000}$

| Dim. | Tol. |
|---|---|
| 1 Pl. dec. | ±.060 |
| 2 Pl. dec. | ±.030 |
| 3 Pl. dec. | ±.005 |

.25 (typ.)

.3 R (typ.)

.19 min. cl.

.2

.25

.31

.1 R

.25 R

.25–20 (4) .5 LG.
3.750 dia. B.C.

2.63 dia.

.25–20–.75 LG.

7.50

1.3

1.000

.675

.25 nom. ball dia.
Brg. pressfit on shaft
slipfit in hsg.

Pressfit in hsg.

.500

Determine

Pitch rad.

2° draft (typ.)

.31

2.31 dia.

.63 R

4.75 dia.
N.T.S.

.38

4.000
N.T.S.

Taper angle of shaft
equals that of hsg.

12–24 int. hex set scr.
.31 LG.

.25 wide flat
1″ long

.373
.374

1.000

.625

.18

1.125

.1 R (typ.)

.1 R (typ.)

**145**

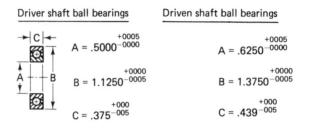

Driver shaft ball bearings

$A = .5000^{+0005}_{-0000}$

$B = 1.1250^{+0000}_{-0005}$

$C = .375^{+000}_{-005}$

Driven shaft ball bearings

$A = .6250^{+0005}_{-0000}$

$B = 1.3750^{+0000}_{-0005}$

$C = .439^{+000}_{-005}$

Figure 9.17
Boston Gear, Quincy, Massachusetts

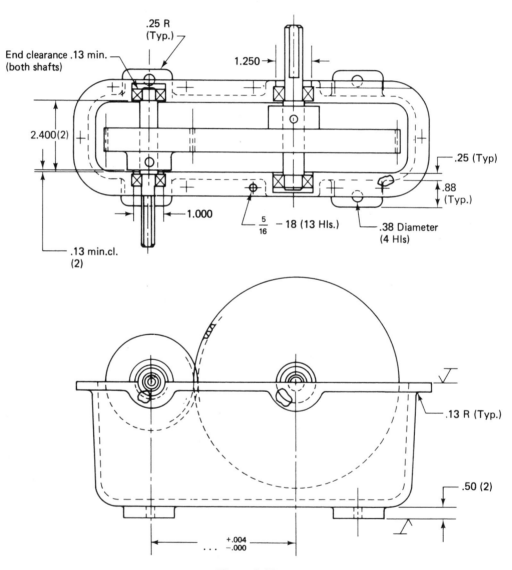

Figure 9.18

146

5. A horizontal shaft supported by sleeve bearings drives a shaft vertically above it by means of a V-belt. Where would you locate an oil groove in the bearing adjacent to the pulley? (a) Near the bottom? (b) Near the top? (c) On one side?

6. A shaft supported by ball bearings has its speed reduced from 775 rpm to 725 rpm and its load reduced from 148 lb to 100 lb. The original rated life of the bearings was 800 h. What will the rated life of the bearings be now?

7. Why is the press fit allowance for ball bearings pressed onto a shaft critical?

8. What are the purposes of seals in bearings with rolling contact? What disadvantage do they have?

# 10

# SHAFTING

*Size Determination. Loads on Bearings. Location of Transmission Elements. Sizes and Recommended Speed Ranges.*

Shafting is used for the transmission of rotation and power and for their distribution by means of pulleys or sheaves (belts), sprockets (chain), or gears.

The main loads on shafts used for the transmission of power are torsion (twist) and bending (see Fig. 10.1). These usually occur in combination. Bending causes deflection which may lead to vibration and whipping at higher revolutions per minute values because of the additional influence of centrifugal force.

Critical items in the design of shafting are the diameter of the shaft, its material, and the location and spacing of bearings and loads. Pulleys, sheaves, sprockets, pinions, etc., should preferably be located as close as possible to the bearings in order to minimize bending loads (Fig. 10.2).

In general, for shafts running at higher speeds, bearings should be more closely spaced and shaft sizes should have a diameter somewhat larger than necessary to prevent whipping and vibration.

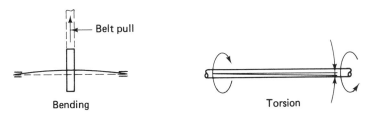

Bending                        Torsion

**Figure 10.1**

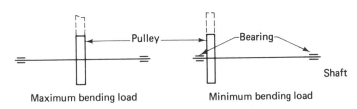

Maximum bending load       Minimum bending load

For the same amount of power transmitted

**Figure 10.2**

Bearings should be properly installed and supported so that lasting alignment is ensured. Misalignment, either initial or occurring in use, may cause eventual breakage of the shaft resulting from fatigue (a type of failure caused by repetitive variations of low stresses).

Table 10.1 shows recommended shaft diameters. These sizes are standardized and are generally commercially available. Standard shafting is made from cold-rolled, low-carbon steel. It has a smooth surface (polished in the larger sizes) and can be freely machined.

For certain industries, special shafting from stainless steel, Monel, bronze, etc., is available to suit chemical (corrosive) or physical conditions, or sanitary requirements prevailing in certain industries. Such special shafting is, of course, more expensive than standard shafting which therefore should be used wherever possible.

If conditions require turning down the end of a shaft, for instance, to fit a pulley or a coupling, it is recommended that as large a fillet as possible be used in order to minimize stress concentrations that may cause fatigue failure (see Fig. 10.3).

Decreases or increases in ambient temperature cause contraction and expansion of shafting. For this reason, only one bearing, the "anchor" bearing, should locate the shaft axially. All other bearings should permit the shaft to move freely in an axial direction (to "float"). The anchor bearing may be located near an important wheel or it may be in the center of the shaft in order to minimize total movement. Several shafts connected

TABLE 10.1

Recommended Shaft Diameters

| | | | | | |
|---|---|---|---|---|---|
| $^{15}/_{16}$ | $1\,^{3}/_{16}$ | $1\,^{7}/_{16}$ | $1\,^{11}/_{16}$ | $1\,^{15}/_{16}$ | $2\,^{3}/_{16}$ |
| $2\,^{7}/_{16}$ | $2\,^{11}/_{16}$ | $2\,^{15}/_{16}$ | $3\,^{7}/_{16}$ | $3\,^{15}/_{16}$ | $4\,^{7}/_{16}$ |
| $4\,^{15}/_{16}$ | $5\,^{7}/_{16}$ | 6 | $6\,^{1}/_{2}$ | 7 | $7\,^{1}/_{2}$ |

Diameter Tolerances—Mild Steel Shafting

| Shaft Size | Oversize | Undersize |
|---|---|---|
| 1″ and under | .000″ | .002″ |
| $1\,^{1}/_{16}-2$ | .000 | .003 |
| $2\,^{1}/_{16}-4$ | .000 | .004 |
| $4\,^{1}/_{16}-6$ | .000 | .005 |

(Courtesy Dodge Div. Reliance Electric Co.)

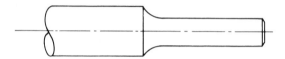

**Figure 10.3**

by solid-type couplings such as flange and sleeve couplings (see Chapter 12) expand and contract as if the shaft were in one piece. If the total expansion is considered excessive, one may use special expansion couplings that absorb changes in length instead of transmitting them. If such couplings are used, each shaft section will need its own anchor bearing.

Table 10.2 gives the lineal expansion of steel shafting in inches (given the shaft length in feet) vs the temperature rise in multiples of 20°F. Intermediate values may be found by interpolation.

Pulleys, gears, couplings, etc., may be fastened to a shaft in several different ways. The most common method is by key (see Fig. 10.4). See Chapter 11 for details on this and other fastening methods.

The *spacing* of bearings on ordinary shafting is usually approximately 8 ft and somewhat more for large-diameter shafts. As mentioned before, it is recommended that closer spacings be used for higher revolutions per minute conditions. The value of 8 ft (see Fig. 10.5) is a recommendation found in many manufacturers' catalogs.

Tables 10.3, 4, 5 & 6 may be used to find guiding values of *recommended diameters* and horsepower under conditions of increasing magnitude of the bending moment going from 10.3 to 10.6 as a function of shaft revolutions per minute (review Chapter 4 if necessary). The values given in these tables are based upon a safe shear stress of 6,000 psi for standard, key-seating shafting.

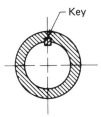

**Figure 10.4**

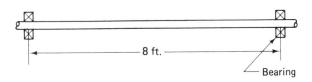

**Figure 10.5**

## TABLE 10.2

*Lineal Expansion of Steel Shafting*
*Based on Expansion in Inches = 0.0000063*
*× 12 × Length in Feet × Temp.*
*Increase in Degrees Fahr.*

| Length Feet | Temperature Increase—Degrees Fahr. | | | | | Length Feet | Temperature Increase—Degrees Fahr. | | | | |
|---|---|---|---|---|---|---|---|---|---|---|---|
| | 20° | 40° | 60° | 80° | 100° | | 20° | 40° | 60° | 80° | 100° |
| 1 | .0015″ | .0030″ | .0045″ | .0060″ | .0075″ | 40 | .060″ | .121″ | .181″ | .242″ | .302″ |
| 2 | .0030 | .0060 | .0091 | .0121 | .0151 | 45 | .068 | .136 | .204 | .272 | .340 |
| 3 | .0045 | .0091 | .0136 | .0181 | .0227 | 50 | .076 | .151 | .227 | .302 | .378 |
| 4 | .0060 | .0121 | .0181 | .0242 | .0302 | 55 | .083 | .166 | .249 | .333 | .416 |
| 5 | .0076 | .0151 | .0227 | .0302 | .0378 | 60 | .091 | .181 | .272 | .363 | .454 |
| 6 | .0091 | .0181 | .0272 | .0363 | .454 | 65 | .098 | .197 | .295 | .393 | .491 |
| 7 | .0106 | .0212 | .0318 | .0423 | .0529 | 70 | .106 | .212 | .317 | .423 | .529 |
| 8 | .0121 | .0242 | .0363 | .0484 | .0605 | 75 | .113 | .227 | .340 | .454 | .567 |
| 9 | .0136 | .0272 | .0408 | .0544 | .0680 | 80 | .121 | .242 | .363 | .484 | .605 |
| 10 | .0151 | .0302 | .0454 | .0605 | .0756 | 85 | .129 | .257 | .386 | .514 | .643 |
| 12 | .0181 | .0363 | .0544 | .0726 | .0907 | 90 | .136 | .272 | .408 | .544 | .680 |
| 14 | .0212 | .0423 | .0635 | .0847 | .1058 | 95 | .144 | .287 | .431 | .575 | .718 |
| 16 | .024 | .048 | .073 | .097 | .121 | 100 | .151 | .302 | .454 | .605 | .756 |
| 18 | .027 | .054 | .082 | .109 | .136 | 110 | .166 | .333 | .499 | .665 | .832 |
| 20 | .030 | .060 | .091 | .121 | .151 | 120 | .181 | .363 | .544 | .726 | .907 |
| 25 | .038 | .076 | .113 | .151 | .189 | 130 | .197 | .393 | .590 | .786 | .983 |
| 30 | .045 | .091 | .136 | .181 | .227 | 140 | .212 | .423 | .635 | .847 | 1.058 |
| 35 | .053 | .106 | .159 | .212 | .265 | 150 | .227 | .454 | .680 | .907 | 1.134 |

(Courtesy: Dodge Div. of Reliance Electric Co.)

## TABLE 10.3

*No Bending Moment. (Shafts without pulleys, sprockets or gears. Torsion only.)*

| Shaft Size | Horse Power at Various Revolutions per Minute | | | | | | | | | | | | | | | | | | |
|---|---|---|---|---|---|---|---|---|---|---|---|---|---|---|---|---|---|---|---|
| | 25 | 50 | 75 | 100 | 125 | 150 | 175 | 200 | 225 | 250 | 275 | 300 | 350 | 400 | 500 | 600 | 700 | 800 | 900 |
| 15/16 | .3 | .7 | 1.1 | 1.5 | 1.9 | 2.3 | 2.6 | 3.0 | 3.4 | 3.8 | 4.2 | 4.6 | 5.3 | 6.1 | 7.7 | 9.2 | 10.7 | 12.3 | 13.8 |
| 1 3/16 | .7 | 1.5 | 2.3 | 3.1 | 3.9 | 4.6 | 5.4 | 6.2 | 7.0 | 7.8 | 8.6 | 9.3 | 10.9 | 12.5 | 15.6 | 18.7 | 21.9 | 25.0 | 28.1 |
| 1 7/16 | 1.3 | 2.7 | 4.1 | 5.5 | 6.9 | 8.3 | 9.7 | 11.1 | 12.4 | 13.8 | 15.2 | 16.6 | 19.4 | 22.2 | 27.7 | 33.3 | 38.8 | 44.4 | 49.9 |
| 1 11/16 | 2.2 | 4.4 | 6.6 | 8.9 | 11.2 | 13.4 | 15.7 | 17.9 | 20.2 | 22.4 | 24.7 | 26.9 | 31.4 | 35.9 | 44.9 | 53.8 | 62.8 | 71.8 | 80.8 |
| 1 5/16 | 3.3 | 6.7 | 10.1 | 13.5 | 16.9 | 20.3 | 23.7 | 27.1 | 30.5 | 33.9 | 37.3 | 40.7 | 47.5 | 54.3 | 67.9 | 81.5 | 95.1 | 108. | 122. |
| 2 3/16 | 4.9 | 9.8 | 14.6 | 19.5 | 24.4 | 29.3 | 34.2 | 39.1 | 44.0 | 48.9 | 53.8 | 58.6 | 68.4 | 78.2 | 97.8 | 117.0 | 136. | 156. | 176. |
| 2 7/16 | 6.7 | 13.5 | 20.2 | 27.0 | 33.8 | 40.6 | 47.3 | 54.1 | 60.9 | 67.6 | 74.4 | 81.2 | 94.7 | 108. | 135. | 162. | 189. | 216. | 243. |
| 2 11/16 | 9.0 | 18.1 | 27.1 | 36.2 | 45.3 | 54.4 | 63.4 | 72.5 | 81.6 | 90.7 | 99.7 | 108. | 126. | 145. | 181. | 217. | 253. | 290. | 326. |
| 2 15/16 | 11.8 | 23.6 | 35.4 | 47.3 | 59.2 | 71. | 82.9 | 94.7 | 106. | 118. | 130. | 142. | 165. | 189. | 236. | 284. | 331. | 379. | 426. |
| 3 7/16 | 19.0 | 37.9 | 57.0 | 75.9 | 94.9 | 113. | 132. | 151. | 170. | 189. | 208. | 227. | 265. | 303. | 379. | 455. | 531. | 607. | 683. |
| 3 15/16 | 28.5 | 57.0 | 85.5 | 114.0 | 142.0 | 171. | 199. | 228. | 256. | 285. | 313. | 342. | 399. | 456. | 570. | 684. | 798. | 912. | 1026. |
| 4 7/16 | 40.8 | 81.6 | 122.0 | 163.0 | 204.0 | 245. | 286. | 327. | 367. | 408. | 449. | 490. | 572. | 653. | 816. | 980. | 1143. | 1306. | 1470. |

(Courtesy: Reliance Electric Co.)

152

# TABLE 10.4

*Limited Bending Moment. (Pulleys, sprockets or gears near bearings. Ordinary line shafts.)*

| Shaft Size | Horse Power at Various Revolutions per Minute | | | | | | | | | | | | | | | | | | |
|---|---|---|---|---|---|---|---|---|---|---|---|---|---|---|---|---|---|---|---|
| | 25 | 50 | 75 | 100 | 125 | 150 | 175 | 200 | 225 | 250 | 275 | 300 | 350 | 400 | 500 | 600 | 700 | 800 | 900 |
| $15/16$ | .2 | .5 | .7 | 1.0 | 1.2 | 1.5 | 1.7 | 2.0 | 2.3 | 2.5 | 2.8 | 3.0 | 3.5 | 4.1 | 5.1 | 6.1 | 7.1 | 8.2 | 9.2 |
| $1\,3/16$ | .5 | 1.0 | 1.5 | 2.0 | 2.6 | 3.1 | 3.6 | 4.1 | 4.7 | 5.2 | 5.7 | 6.2 | 7.3 | 8.3 | 10.4 | 12.5 | 14.6 | 16.7 | 18.8 |
| $1\,7/16$ | .9 | 1.8 | 2.7 | 3.7 | 4.6 | 5.5 | 6.4 | 7.4 | 8.3 | 9.2 | 10.1 | 11.1 | 12.9 | 14.8 | 18.5 | 22.2 | 25.9 | 29.6 | 33.3 |
| $1\,11/16$ | 1.4 | 2.9 | 4.3 | 5.9 | 7.4 | 8.9 | 10.4 | 11.9 | 13.4 | 14.9 | 16.4 | 17.9 | 20.9 | 23.9 | 29.9 | 35.9 | 41.9 | 47.9 | 53.9 |
| $1\,15/16$ | 2.2 | 4.5 | 6.7 | 9.0 | 11.3 | 13.6 | 15.8 | 18.1 | 20.4 | 22.6 | 24.9 | 27.2 | 31.7 | 36.2 | 45.3 | 54.4 | 63.4 | 72.5 | 81.6 |
| $2\,3/16$ | 3.2 | 6.5 | 9.7 | 13.0 | 16.3 | 19.5 | 22.8 | 26.1 | 29.3 | 32.6 | 35.8 | 39.1 | 45.6 | 52.2 | 65.2 | 78.3 | 91.3 | 104. | 117. |
| $2\,7/16$ | 4.5 | 9.0 | 13.5 | 18.0 | 22.5 | 27.0 | 31.6 | 36.1 | 40.6 | 45.1 | 49.6 | 54.1 | 63.2 | 72.2 | 90.2 | 108. | 126. | 144. | 162. |
| $2\,11/16$ | 6.0 | 12.1 | 18.1 | 24.2 | 30.2 | 36.3 | 42.3 | 48.4 | 54.4 | 60.5 | 66.5 | 72.6 | 84.7 | 96.8 | 121. | 145. | 169. | 193. | 217. |
| $2\,15/16$ | 7.9 | 15.8 | 23.7 | 31.6 | 39.5 | 47.4 | 55.3 | 63.2 | 71.1 | 79.0 | 86.9 | 94.8 | 110. | 126. | 158. | 189. | 221. | 252. | 284. |
| $3\,7/16$ | 12.6 | 25.3 | 37.9 | 50.6 | 63.3 | 75.9 | 88.6 | 101. | 113. | 126. | 139. | 151. | 177. | 202. | 253. | 303. | 354. | 405. | 455. |
| $3\,15/16$ | 19.0 | 38.0 | 57.0 | 76.1 | 94.1 | 114. | 133. | 152. | 171. | 190. | 209. | 228. | 266. | 304. | 380. | 456. | 532. | 608. | 685. |
| $4\,7/16$ | 27. | 54. | 81. | 108. | 136. | 163. | 190. | 217. | 245. | 272. | 299. | 326. | 381. | 435. | 544. | 653. | 762. | 871. | 980. |
| $4\,15/16$ | 37. | 75. | 112. | 150. | 187. | 225. | 262. | 300. | 337. | 375. | 412. | 450. | 525. | 600. | 750. | 900. | 1050. | 1200. | 1350. |
| $5\,7/16$ | 50. | 100. | 150. | 200. | 250. | 300. | 350. | 400. | 451. | 501. | 551. | 601. | 701. | 801. | 1002. | 1202. | 1403. | 1603. | 1804. |
| $5\,15/16$ | 65. | 130. | 195. | 261. | 326. | 391. | 456. | 522. | 587. | 652. | 717. | 783. | 913. | 1044. | 1305. | 1566. | 1827. | 2088. | 2349. |
| $6\,1/2$ | 85. | 171. | 256. | 342. | 427. | 513. | 598. | 684. | 769. | 855. | 940. | 1026. | 1197. | 1368. | 1710. | 2052. | 2394. | 2736. | 3078. |

(Courtesy: Reliance Electric Co.)

# TABLE 10.5

### Heavy Bending Moment. (*Use for main or important shafts.*)

| Shaft Size | Horse Power at Various Revolutions per Minute | | | | | | | | | | | | | | | | | | |
|---|---|---|---|---|---|---|---|---|---|---|---|---|---|---|---|---|---|---|---|
| | 25 | 50 | 75 | 100 | 125 | 150 | 175 | 200 | 225 | 250 | 275 | 300 | 350 | 400 | 500 | 600 | 700 | 800 | 900 |
| 1 11/16 | .8 | 1.7 | 2.5 | 3.5 | 4.4 | 5.3 | 6.2 | 7.1 | 8.0 | 8.9 | 9.8 | 10.7 | 12.5 | 14.3 | 17.9 | 21.5 | 25.1 | 28.7 | 32.3 |
| 1 15/16 | 1.3 | 2.7 | 4.0 | 5.4 | 6.7 | 8.1 | 9.5 | 10.8 | 12.2 | 13.5 | 14.9 | 16.3 | 19.0 | 21.7 | 27.1 | 32.6 | 38.0 | 43.5 | 48.9 |
| 2 3/16 | 1.9 | 3.9 | 5.8 | 7.8 | 9.7 | 11.7 | 13.7 | 15.6 | 17.6 | 19.5 | 21.5 | 23.4 | 27.4 | 31.3 | 39.1 | 46.9 | 54.8 | 62.6 | 70.4 |
| 2 7/16 | 2.7 | 5.4 | 8.1 | 10.8 | 13.5 | 16.2 | 18.9 | 21.6 | 24.3 | 27.0 | 29.7 | 32.4 | 37.9 | 43.3 | 54.1 | 64.9 | 75.8 | 86.6 | 97.4 |
| 2 11/16 | 3.6 | 7.2 | 10.8 | 14.5 | 18.1 | 21.7 | 25.4 | 29.0 | 32.6 | 36.2 | 39.9 | 43.5 | 50.8 | 58.0 | 72.5 | 87.1 | 101. | 116. | 130. |
| 2 15/16 | 4.7 | 9.4 | 14.1 | 18.9 | 23.6 | 28.4 | 33.1 | 37.9 | 42.6 | 47.3 | 52.1 | 56.8 | 66.3 | 75.8 | 94.7 | 113. | 132. | 151. | 170. |
| 3 7/16 | 7.5 | 15.1 | 22.6 | 30.3 | 37.9 | 45.5 | 53.1 | 60.7 | 68.3 | 75.9 | 83.5 | 91.1 | 106. | 121. | 151. | 182. | 212. | 243. | 273. |
| 3 15/16 | 11.4 | 22.8 | 34.2 | 45.6 | 57.0 | 68.4 | 79.9 | 91.3 | 102. | 114. | 125. | 136. | 159. | 182. | 228. | 273. | 319. | 365. | 410. |
| 4 7/16 | 16.3 | 32.6 | 48.9 | 65.3 | 81.6 | 98.0 | 114. | 130. | 147. | 163. | 179. | 196. | 228. | 261. | 326. | 392. | 457. | 522. | 588. |
| 4 15/16 | 22.5 | 45.0 | 67.5 | 90.0 | 112. | 135. | 157. | 180. | 202. | 225. | 247. | 270. | 315. | 360. | 450. | 540. | 630. | 720. | 810. |
| 5 7/16 | 30.0 | 60.0 | 90.0 | 120. | 150. | 180. | 210. | 240. | 270. | 300. | 330. | 360. | 420. | 480. | 601. | 721. | 841. | 961. | 1082. |
| 5 15/16 | 39.0 | 78.0 | 117. | 156. | 195. | 234. | 273. | 313. | 352. | 391. | 430. | 469. | 547. | 626. | 782. | 939. | 1095. | 1252. | 1409. |
| 6 1/2 | 51.0 | 102. | 153. | 205. | 256. | 308. | 359. | 410. | 462. | 513. | 564. | 616. | 718. | 821. | 1027. | 1232. | 1437. | 1643. | 1848. |
| 7 | 64.0 | 128. | 192. | 256. | 320. | 384. | 448. | 513. | 577. | 641. | 705. | 769. | 897. | 1026. | 1282. | 1539. | 1795. | 2052. | 2308. |
| 7 1/2 | 78.5 | 157. | 235. | 315. | 394. | 473. | 552. | 631. | 709. | 788. | 867. | 946. | 1104. | 1262. | 1577. | 1893. | 2208. | 2524. | 2839. |
| 8 | 95.5 | 191. | 286. | 382. | 478. | 574. | 670. | 765. | 861. | 957. | 1053. | 1148. | 1340. | 1531. | 1914. | 2297. | 2680. | 3063. | 3446. |
| 8 1/2 | 114. | 229. | 343. | 459. | 574. | 688. | 803. | 918. | 1033. | 1148. | 1263. | 1377. | 1607. | 1837. | 2296. | 2755. | 3215. | 3674. | 4133. |
| 9 | 136. | 272. | 408. | 545. | 681. | 817. | 954. | 1090. | 1226. | 1363. | 1499. | 1635. | 1908. | 2181. | 2726. | 3271. | 3816. | 4362. | 4907. |
| 9 1/2 | 160. | 320. | 480. | 641. | 801. | 961. | 1122. | 1282. | 1442. | 1603. | 1763. | 1923. | 2244. | 2565. | 3206. | 3847. | 4488. | 5130. | 5771. |
| 10 | 186. | 373. | 559. | 747. | 934. | 1121. | 1308. | 1495. | 1682. | 1869. | 2056. | 2243. | 2617. | 2991. | 3739. | 4487. | 5235. | 5893. | 6731. |

# TABLE 10.6

*Severe Conditions. (Heavy shock loads. Excessively tight belts. Long clutch sleeves.)*

| Shaft Size | Horse Power at Various Revolutions per Minute | | | | | | | | | | | | | | | | | | |
|---|---|---|---|---|---|---|---|---|---|---|---|---|---|---|---|---|---|---|---|
| | 25 | 50 | 75 | 100 | 125 | 150 | 175 | 200 | 225 | 250 | 275 | 300 | 350 | 400 | 500 | 600 | 700 | 800 | 900 |
| $1\frac{11}{16}$ | .4 | .8 | 1.2 | 1.7 | 2.2 | 2.6 | 3.1 | 3.5 | 4.0 | 4.4 | 4.9 | 5.3 | 6.2 | 7.1 | 8.9 | 10.7 | 12.5 | 14.3 | 16.1 |
| $1\frac{15}{16}$ | .6 | 1.3 | 2.0 | 2.7 | 3.3 | 4.0 | 4.7 | 5.4 | 6.1 | 6.7 | 7.4 | 8.1 | 9.5 | 10.8 | 13.5 | 16.3 | 19.0 | 21.7 | 24.4 |
| $2\frac{3}{16}$ | .9 | 1.9 | 2.9 | 3.9 | 4.8 | 5.8 | 6.8 | 7.8 | 8.8 | 9.7 | 10.7 | 11.7 | 13.7 | 15.6 | 19.5 | 23.4 | 27.4 | 31.3 | 35.2 |
| $2\frac{7}{16}$ | 1.3 | 2.7 | 4.0 | 5.4 | 6.7 | 8.1 | 9.4 | 10.8 | 12.1 | 13.5 | 14.8 | 16.2 | 18.9 | 21.6 | 27.0 | 32.4 | 37.9 | 43.3 | 48.7 |
| $2\frac{11}{16}$ | 1.8 | 3.6 | 5.4 | 7.2 | 9.0 | 10.8 | 12.7 | 14.5 | 16.3 | 18.1 | 19.9 | 21.7 | 25.4 | 29.0 | 36.2 | 43.5 | 50.5 | 58.0 | 65.0 |
| $2\frac{15}{16}$ | 2.3 | 4.7 | 7.0 | 9.4 | 11.8 | 14.2 | 16.5 | 18.9 | 21.3 | 23.6 | 26.0 | 28.4 | 33.1 | 37.9 | 47.3 | 56.5 | 66.0 | 75.5 | 85.0 |
| $3\frac{7}{16}$ | 3.7 | 7.5 | 11.3 | 15.1 | 18.9 | 22.7 | 26.5 | 30.3 | 34.1 | 37.9 | 41.7 | 45.5 | 53.0 | 60.5 | 75.5 | 91.0 | 106. | 121. | 136. |
| $3\frac{15}{16}$ | 5.7 | 11.4 | 17.1 | 22.8 | 28.5 | 34.2 | 39.9 | 45.6 | 51.0 | 57.0 | 62.5 | 68.0 | 79.5 | 91.0 | 114. | 136. | 159. | 182. | 205. |
| $4\frac{7}{16}$ | 8.1 | 16.3 | 24.4 | 32.6 | 40.8 | 49.0 | 57.0 | 65.0 | 73.5 | 81.5 | 89.5 | 98.0 | 114. | 130. | 163. | 196. | 228. | 261. | 294. |
| $4\frac{15}{16}$ | 11.2 | 22.5 | 33.7 | 45.0 | 56.0 | 67.5 | 78.5 | 90.0 | 101. | 112. | 123. | 135. | 157. | 180. | 225. | 270. | 315. | 360. | 405. |
| $5\frac{7}{16}$ | 15.0 | 30.0 | 45.0 | 60.0 | 75.0 | 90.0 | 105. | 120. | 135. | 150. | 165. | 180. | 210. | 240. | 300. | 360. | 420. | 480. | 541. |
| $5\frac{15}{16}$ | 19.5 | 39.0 | 58.5 | 78.0 | 97.5 | 117. | 136. | 156. | 171. | 195. | 215. | 234. | 273. | 313. | 391. | 469. | 547. | 626. | 704. |
| $6\frac{1}{2}$ | 25.5 | 51.0 | 76.5 | 102.5 | 128. | 154. | 179. | 205. | 231. | 256. | 282. | 308. | 359. | 410. | 513. | 616. | 718. | 821. | 924. |
| 7 | 32.0 | 64.0 | 96.0 | 128.0 | 160. | 192. | 224. | 256. | 288. | 320. | 352. | 384. | 448. | 513. | 641. | 769. | 897. | 1026. | 1154. |
| $7\frac{1}{2}$ | 39.2 | 78.5 | 117. | 157. | 197. | 236. | 276. | 315. | 354. | 394. | 433. | 473. | 552. | 631. | 788. | 946. | 1104. | 1262. | 1419. |
| 8 | 47.7 | 95.5 | 143. | 191. | 239. | 287. | 335. | 382. | 430. | 478. | 526. | 574. | 670. | 765. | 957. | 1148. | 1340. | 1531. | 1723. |
| $8\frac{1}{2}$ | 57.0 | 114. | 171. | 229. | 287. | 344. | 401. | 459. | 516. | 574. | 631. | 688. | 803. | 918. | 1148. | 1377. | 1607. | 1837. | 2066. |
| 9 | 68.0 | 136. | 204. | 272. | 340. | 408. | 477. | 545. | 613. | 681. | 749. | 817. | 954. | 1090. | 1363. | 1635. | 1908. | 2181. | 2453. |
| $9\frac{1}{2}$ | 80.0 | 160. | 240. | 320. | 400. | 480. | 561. | 641. | 721. | 801. | 881. | 961. | 1122. | 1282. | 1603. | 1923. | 2244. | 2565. | 2885. |
| 10 | 93.0 | 186. | 279. | 373. | 467. | 560. | 654. | 747. | 841. | 934. | 1028. | 1121. | 1308. | 1495. | 1869. | 2243. | 2617. | 2991. | 3365. |

Caution—Be generous in the selection of shaft diameters as liberal diameters not only reduce deflection and vibration but also generally increase bearing life.
(Courtesy: Reliance Electric Co.)

155

It should be noted that the *stiffness* of shafting, i.e., its ability to resist bending, is not affected by the strength of the steel used. However, the stiffness is proportional to the cube of the diameter, which is the reason why a somewhat larger diameter than that found in the tables should be used for conditions of higher revolutions per minute.

An example showing how to use these tables is given below.

**Given:** A shaft is to transmit 25 hp at 500 rpm. All pulleys are located near bearings. The shaft is 40 ft long and is made up of two 20-ft sections. The anchor bearing is near the center. The temperature variation is 100° F.

**Required:** Recommended standard shaft diameter, spacing of bearings, and thermal variation of length at the ends of the shaft.

**Solution:** Go to Table 10.4 for limited bending moment. Find 500 rpm (fifth column from right). Go down vertically to the nearest horsepower value larger than 25, which is 29.9. Then go left horizontally and find shaft size of $1^{11}/_{16}$ in. diameter. Since the revolutions per minute are fairly high, use a spacing of approximately 7 ft. A total of seven bearings is required. Since the anchor bearing is near the center of the shaft, the maximum shaft length affected by temperature changes is 20 ft on either side of the anchor bearing. Go to Table 10.2. Find 20 ft in the left column and then move right to the column labeled 100° F and read .151 in. total change of length at each end of the 20 ft of shafting.

Shafts are sometimes suspended from ceilings or mounted on overhead beams. In such cases, *pillow blocks* are used for support.

Pillow blocks may be equipped with journal (sleeve) bearings when used with light loads or they may be equipped with bearings with rolling contact such as ball and roller bearings when used with heavier loads.

A great many different types of pillow block exist to suit various industrial requirements. Most pillow blocks have self-aligning bearings, which simplify installation because they do not require great angular precision.

When shafts pass through vertical walls, *flange blocks* are used. Flange blocks use the same types of bearing as pillow blocks.

An *anchor block* with a journal bearing may locate the shaft by locking collars equipped with setscrews on each side. When the anchor block is equipped with either ball bearings or roller bearings, the inner race usually has one or more setscrews. These may be used either to lock the inner race to the shaft in the case of an anchor block or to cause the inner race to rotate with the shaft while permitting axial motion. In the latter case, a groove or grooves are required in the shaft in way of the setscrews.

Journal bearing-type pillow or flange blocks may be self-lubricating (sintered bronze) or they may be equipped with oil cups and felt-packed reservoirs. After the reservoir has been filled, these bearings usually operate for a long time without having to have oil added.

Industrial-type mineral oils or plain (nondetergent) automotive crankcase oils are recommended.

Pillow or flange blocks with rolling contact bearings are usually grease-lubricated. Some are lubricated for life and others must be re-lubricated (they have grease fittings for that purpose).

See Chapter 9 for more details on the bearings described above.

## 10.1 FORCES ON BEARINGS FOR SHAFTING

A certain amount of tension in a belt or rope is necessary to keep slipping and/or creep at an acceptable value.

The amount of preloading (initial tension) required depends on the type of belt and the belt material used. Flat belts generally need more tension than V-belts. Chain needs no initial tension to operate properly; toothed belts ("timing belts") need only a nominal amount.

To approximate the radial force exerted by a pulley or sprocket on the *shaft* (not the bearing), the following equation may be used:

$$K_r = f_1 f_d K_p$$

For belts with exceptionally large cross sections, the following equation may be used:

$$K_r = f_2 A$$

Where $K_r$ = radial force on the shaft, $K_p$ = effective belt, rope, or chain pull (see page 158), $f_1$, $f_2$, and $f_d$ are factors found in Table 10.7 and $A$ = cross-sectional area of the rope or belt in square inches.

TABLE 10.7

*Factors for Belt, Rope, and Chain Drive Calculations*

| Type of Drive | $f_1$ | $f_d$ | $f_2$ |
|---|---|---|---|
| Flat leather belt with tension pulley . . . . . . . . . | 1.75 to 2.5 | 1 to 1.1 | 550 |
| Flat leather belt without tension pulley . . . . . .<br>Fabric belt, rubberized canvas belt . . . . . . . . . .<br>Balata belt. . . . . . . . . . . . . . . . . . . . . . . . . . . | 2.25 to 3.5 | 1 to 1.2 | 800 |
| V-belt. . . . . . . . . . . . . . . . . . . . . . . . . . . . . . . | 1.5 to 2 | 1 to 1.2 | 275 |
| Steel belt . . . . . . . . . . . . . . . . . . . . . . . . . . | 4 to 6 | 1 to 1.2 | |
| Cotton or hemp rope . . . . . . . . . . . . . . . . . . | 2 to 6 | 1 to 1.2 | 700 |
| Chain . . . . . . . . . . . . . . . . . . . . . . . . . . . . . . | 1 | 1.1 to 1.5 | |

(Courtesy: SKF Engineering Data.)

The value of $K_p$ may be calculated from

$$K_p = \frac{\text{hp} \times 126{,}000}{\text{rpm} \times D_p}$$

Where hp = transmitted horsepower, rpm = revolutions per minute of the shaft, and $D_p$ = pitch diameter of the pulley or sprocket.

It should be noted that these equations ignore the effects of the weight of the belt or rope.

The computation of *bearing loads* on long shafting with several pulleys is in general complicated and outside the scope of this text. These loads depend on the magnitude and direction of various effective belt (or rope or chain) pulling forces ($K_p$), the number and location of the bearings, and the combined weight of the shaft, pulleys, and belts.

For a single sheave or sprocket, the computation is easier and may be done with the aid of the equations shown in Fig. 10.6 for an *overhung hub load* (sheave located outside the bearings) and in Fig. 10.7 for *outboard bearings* (sheave between the bearings). The hub load in these equations is the value of $K_r$ in the equations on page 157.

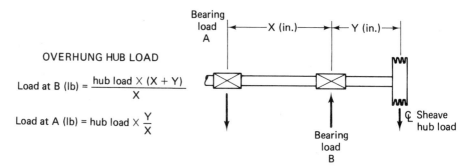

OVERHUNG HUB LOAD

Load at B (lb) = $\dfrac{\text{hub load} \times (X + Y)}{X}$

Load at A (lb) = hub load $\times \dfrac{Y}{X}$

**Figure 10.6**
Eaton Corporation, Cleveland, Ohio

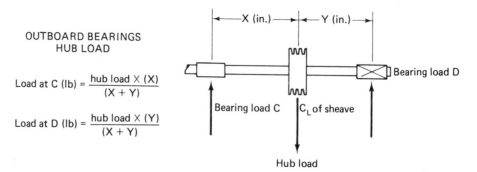

OUTBOARD BEARINGS
HUB LOAD

Load at C (lb) = $\dfrac{\text{hub load} \times (X)}{(X + Y)}$

Load at D (lb) = $\dfrac{\text{hub load} \times (Y)}{(X + Y)}$

**Figure 10.7**
Eaton Corporation, Cleveland, Ohio

Manufacturers' catalogs for pillow blocks and flange blocks usually give the recommended maximum radial load ratings for a range of revolutions per minute values and shaft sizes.

Below is an example of how to calculate bearing forces.

**Given:** A $2\,{}^3/_{16}$-diameter shaft is supported by two bearings 34 in. on centers. An overhung, multiple V-belt pulley of 12 in.-pitch diameter has the center of belt pull 7.5 in. away from the center of the nearest bearing. The shaft transmits 15 hp at 650 rpm.

**Required:** The loads on each bearing (see Fig. 10.8).

**Solution:** Find the effective belt pull ($K_p$)

$$K_p = \frac{\text{hp} \times 126{,}000}{\text{rpm} \times D_p}$$

$$= \frac{15 \times 126{,}000}{650 \times 12}$$

$$= 242\ \text{lb}$$

Enter this value in $K_r$

$$= f_1 f_d K_p \quad (\text{p. 157})$$

$$= 1.5 \times 1 \times 242 = 363\ \text{lb}$$

The load on the bearing farthest from the pulley is

$$F_A = 363 \times \frac{7.5}{34} = 80\ \text{lb}\quad (\text{p. 158})$$

The load on the bearing nearest to the pulley is

$$F_B = 363 \times \frac{(34 + 7.5)}{34} = 443\ \text{lb}$$

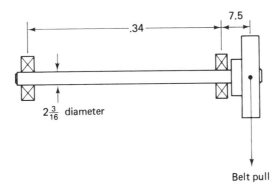

**Figure 10.8**

## PROBLEMS

1. In an existing shaft installation of $1^{11}/_{16}$ in. diameter and with a service speed of 470 rpm, a pulley must be added approximately in the center of the distance between two bearings. The shaft size is adequate to handle the additional horsepower load imposed. The existing bearing spacing is 7'6 in. What modifications, if any, would you suggest for this installation to ensure continued satisfactory operation?

2. A shaft designed for the transmission of 56 hp at 700 rpm is found to be vibrating at that speed and is therefore slowed down to 600 rpm, at which point vibration no longer is a problem. What effect does this have on the shaft's horsepower transmission when the torque remains constant?

3. A short shaft transmitting 5 hp and subjected to torsion only (without pulleys, sprockets, or pinions) has its speed doubled. What happens to its horsepower transmission capability? (Check tables.) Can you explain your answer?

4. A chain sprocket of 9-in. $D_p$ is located between two bearings $A$ and $B$. Its center is 14.5 in. away from the center of bearing $A$ and 19 in. away from the center of bearing $B$. It transmits 7 hp at 475 rpm. What are the loads on the bearings?

5. A shaft is to run at 260 rpm and must transmit 130 hp. Select a diameter when there is no bending load.

**11**

# FASTENING TORQUE- TRANSMITTING ELEMENTS

*Pulleys, Sprockets, Gears, Couplings, Cranks.*

Several methods are used to connect torque-transmitting elements to a shaft. Their purpose is to prevent relative motion between shaft and elements in the direction of rotation, and in most cases also in an axial direction.

For fractional horsepower applications, *setscrews* are often used (see Fig. 11.1). These are headless screws that have an internal hexagon socket at one end and a mostly truncated conical tip at the other end. They are threaded into the hub of the pulley, gear, etc., and when they are seated, they should not protrude outside the hub's surface. The point of contact of the shaft may be in the shape of a conical depression, a flat surface, or a groove (keyway).

In some cases, two setscrews are used either opposite one another or at right angles to one another (see Fig. 11.2). Opposite setscrews cancel each other's induced friction and should not be used. Those at right angles

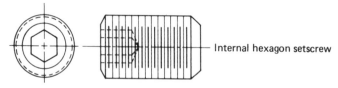

Internal hexagon setscrew

Figure 11.1

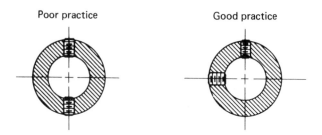

Poor practice          Good practice

Figure 11.2

act in the same general direction and thus provide increased torsional holding strength.

A somewhat stronger connection is obtained by *radial pins.* There are two categories of radial pins: taper pins and cylindrical pins.

*Taper pins* have a standard taper of ½ in. per foot. They are inexpensive and effective, but they require special tapered reamers to provide proper seating and locking. Using a tapered pin in a drilled, cylindrical hole is to be condemned because it is conducive to starting shock loads which may lead to eventual loosening and breakage of the pin.

Taper pins usually pass through the axis of the shaft, but sometimes they are located eccentrically. The latter method provides less weakening of the shaft but the eccentric hole is more expensive to make. Figure 11.3 shows both methods.

*Cylindrical pins* (solid) have the drawback that they are difficult to drive in because they need an interference fit to stay in place. Some modifications have been made to overcome this drawback:

*Groove pins* have one or more longitudinal grooves over part of, or most of, the length. These grooves are the result of a staking (cold forging) operation that results in a slight radial upset. When the pin is driven, the upset material locks in the hole (see Fig. 11.4).

*Spring pins* or *roll pins* are essentially a hollow tube that has tapered ends and a longitudinal slot (see Fig. 11.5). This construction permits the diameter to be somewhat reduced by spring action. The tapered end facilitates driving the pin into a slightly smaller hole than the pin's free

162

Figure 11.3

Figure 11.4

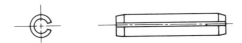

Figure 11.5

diameter, which locks it in place. Spring pins may be removed and reused many times.

Since roll pins are made in different diameters such that they make a sliding fit into one another, it is possible to increase the strength of a roll pin assembly by driving the next smaller size into the first, and so on. These multiple assemblies have a high resistance to fatigue failure.

Single roll pin assemblies have the capability to absorb shock loads. They are economical to use because they work well in a drilled hole only and no reaming is necessary.

*Spiral pins* have an action similar to that of spring pins, but they are not as flexible (see Fig. 11.6).

The most generally used method of fastening torque-transmitting elements to a shaft is by *key and keyway* (see Fig. 11.7). This method is suitable from fractional horsepower applications up to the highest transmitted torques.

The simplest shape of key has a square section that penetrates the shaft and hub about equally. It is sometimes tapered in a radial direction to facilitate assembly. Standard ASA square keys start at $1/8$ in. Suggested sizes for given shaft diameter ranges are given in Table 11.1. A setscrew is most often used to lock the hub axially.

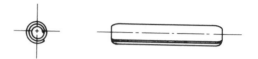

Figure 11.6

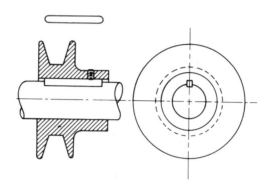

Pulley fastened by
square key and setscrew

Figure 11.7

TABLE 11.1

| Shaft diameter (in.) | Width and height of key (in.) |
| --- | --- |
| $1/2 - 9/16$ | $1/8$ |
| $5/8 - 7/8$ | $3/16$ |
| $15/16 - 1 1/4$ | $1/4$ |
| $1 5/16 - 1 3/8$ | $5/16$ |
| $1 7/16 - 1 3/4$ | $3/8$ |
| $1 13/16 - 2 1/4$ | $1/2$ |
| $2 5/16 - 2 3/4$ | $5/8$ |
| $2 7/8 - 3 1/4$ | $3/4$ |
| $3 3/8 - 3 3/4$ | $7/8$ |
| $3 7/8 - 4 1/2$ | $1$ |
| $4 3/4 - 5 1/2$ | $1 1/4$ |
| $5 3/4 - 6$ | $1 1/2$ |

(Courtesy: *Mark's Handbook*, McGraw-Hill, New York.)

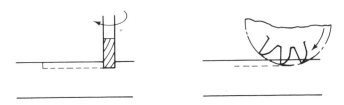

Figure 11.8

The minimum suggested length of keys is 4 times the width and the maximum recommended length is 16 times the width.

While the groove in the hub is usually broached and continuous from one side to the other, there are two ways of milling the groove into the shaft, as illustrated in Fig. 11.8. The method on the left, which utilizes an end mill, produces a keyway that can be filled completely by a key having rounded ends (see Fig. 11.7).

Where it is desired to slide the hub along the shaft, as for instance for the engagement or disengagement of a gear, a *feather key* is often used. A feather key is a continuous square key that is usually fastened to the shaft. The mating hub keyway is made a sliding fit, thus permitting axial motion over the length of the key, while the hub remains locked torsionally.

Square keys have a tendency to tip under heavy loads. This drawback is reduced by using keys of rectangular cross section whose smallest dimension is placed in a radial direction. Such keys are called *flat keys* and are often used in machine construction. *Woodruff keys* (Fig. 11.9) are in the form of a circle segment and were designed to facilitate removal of the keyed part from the shaft. They have the additional advantage of being much less prone to tipping because they penetrate the shaft much more deeply. For long hubs, two Woodruff keys should be used. Standard Woodruff key dimensions are given in Table 11.2.

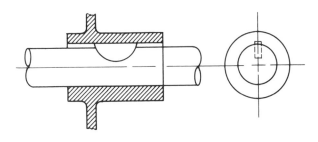

Woodruff key
observe great penetration

Figure 11.9

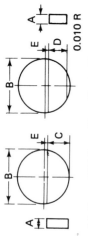

McGraw-Hill Inc., New York, New York

## TABLE 11.2
### Woodruff Key Dimensions

| Key No. | Nominal key size, A × B | Width of key, A Max | Width of key, A Min | Diam of key, B Max | Diam of key, B Min | Height of key C Max | Height of key C Min | Height of key D Max | Height of key D Min | Distance below center, E |
|---|---|---|---|---|---|---|---|---|---|---|
| 204 | $1/16 \times 1/2$ | 0.0635 | 0.0625 | 0.500 | 0.490 | 0.203 | 0.198 | 0.194 | 0.188 | $3/64$ |
| 304 | $3/32 \times 1/2$ | 0.0948 | 0.0928 | 0.500 | 0.490 | 0.203 | 0.198 | 0.194 | 0.188 | $3/64$ |
| 305 | $3/32 \times 5/8$ | 0.0948 | 0.0938 | 0.625 | 0.615 | 0.250 | 0.245 | 0.240 | 0.234 | $1/16$ |
| 404 | $1/8 \times 1/2$ | 0.1260 | 0.1250 | 0.500 | 0.490 | 0.203 | 0.198 | 0.194 | 0.188 | $3/64$ |
| 405 | $1/8 \times 5/8$ | 0.1260 | 0.1250 | 0.625 | 0.615 | 0.250 | 0.245 | 0.240 | 0.234 | $1/16$ |
| 406 | $1/8 \times 3/4$ | 0.1260 | 0.1250 | 0.750 | 0.740 | 0.313 | 0.308 | 0.303 | 0.297 | $1/16$ |
| 505 | $5/32 \times 5/8$ | 0.1573 | 0.1563 | 0.625 | 0.615 | 0.250 | 0.245 | 0.240 | 0.234 | $1/16$ |
| 506 | $5/32 \times 3/4$ | 0.1573 | 0.1563 | 0.750 | 0.740 | 0.313 | 0.308 | 0.303 | 0.297 | $1/16$ |
| 507 | $5/32 \times 7/8$ | 0.1573 | 0.1563 | 0.875 | 0.865 | 0.375 | 0.370 | 0.365 | 0.359 | $1/16$ |
| 606 | $3/16 \times 3/4$ | 0.1885 | 0.1875 | 0.750 | 0.740 | 0.313 | 0.308 | 0.303 | 0.297 | $1/16$ |
| 607 | $3/16 \times 7/8$ | 0.1885 | 0.1875 | 0.875 | 0.865 | 0.375 | 0.370 | 0.365 | 0.359 | $1/16$ |
| 608 | $3/16 \times 1$ | 0.1885 | 0.1875 | 1.000 | 0.990 | 0.438 | 0.433 | 0.428 | 0.422 | $1/16$ |
| 609 | $3/16 \times 1\,1/8$ | 0.1885 | 0.1875 | 1.125 | 1.115 | 0.484 | 0.479 | 0.475 | 0.469 | $5/64$ |

0.010 R

## TABLE 11.2 (cont.)
### Woodruff Key Dimensions

| Key No. | Nominal key size, A × B | Width of key, A Max | Width of key, A Min | Diam of key, B Max | Diam of key, B Min | Height of key C Max | Height of key C Min | D Max | D Min | Distance below center, E |
|---|---|---|---|---|---|---|---|---|---|---|
| 807 | $1/4 \times 7/8$ | 0.2510 | 0.2500 | 0.875 | 0.865 | 0.375 | 0.370 | 0.365 | 0.359 | $1/16$ |
| 808 | $1/4 \times 1$ | 0.2510 | 0.2500 | 1.000 | 0.990 | 0.438 | 0.433 | 0.428 | 0.422 | $1/16$ |
| 809 | $1/4 \times 1\,1/8$ | 0.2510 | 0.2500 | 1.125 | 1.115 | 0.484 | 0.479 | 0.475 | 0.469 | $5/64$ |
| 810 | $1/4 \times 1\,1/4$ | 0.2510 | 0.2500 | 1.250 | 1.240 | 0.547 | 0.542 | 0.537 | 0.531 | $5/64$ |
| 811 | $1/4 \times 1\,3/8$ | 0.2510 | 0.2500 | 1.375 | 1.365 | 0.594 | 0.589 | 0.584 | 0.578 | $3/32$ |
| 812 | $1/4 \times 1\,1/2$ | 0.2510 | 0.2500 | 1.500 | 1.490 | 0.641 | 0.636 | 0.631 | 0.625 | $7/64$ |
| 1008 | $5/16 \times 1$ | 0.3135 | 0.3125 | 1.000 | 0.990 | 0.438 | 0.433 | 0.428 | 0.422 | $1/16$ |
| 1009 | $5/16 \times 1\,1/8$ | 0.3135 | 0.3125 | 1.125 | 1.115 | 0.484 | 0.479 | 0.475 | 0.469 | $5/64$ |
| 1010 | $5/16 \times 1\,1/4$ | 0.3135 | 0.3125 | 1.250 | 1.240 | 0.547 | 0.542 | 0.537 | 0.531 | $5/64$ |
| 1011 | $5/16 \times 1\,3/8$ | 0.3135 | 0.3125 | 1.375 | 1.365 | 0.594 | 0.589 | 0.584 | 0.578 | $3/32$ |
| 1012 | $5/16 \times 1\,1/2$ | 0.3135 | 0.3125 | 1.500 | 1.490 | 0.641 | 0.636 | 0.631 | 0.625 | $7/64$ |
| 1210 | $3/8 \times 1\,1/4$ | 0.3760 | 0.3750 | 1.250 | 1.240 | 0.547 | 0.542 | 0.537 | 0.531 | $5/64$ |
| 1211 | $3/8 \times 1\,3/8$ | 0.3760 | 0.3750 | 1.375 | 1.365 | 0.594 | 0.589 | 0.584 | 0.578 | $3/32$ |
| 1212 | $3/8 \times 1\,1/2$ | 0.3760 | 0.3750 | 1.500 | 1.490 | 0.641 | 0.636 | 0.631 | 0.625 | $7/64$ |

Numbers indicate the nominal key dimensions. The last two digits give the nominal diameter ($B$) in eighths of an inch and the digits preceding the last two give the nominal width ($A$) in thirty-seconds of an inch. Thus, 204 indicates a key $2/32 \times 4/8$ or $1/16 \times 1/2$ in.; 1210 indicates a key $12/32 \times 10/8$ or $3/8 \times 1\,1/4$ in.

(Courtesy: *Mark's Handbook*, McGraw-Hill, NY.)

Square keys, flat keys, and Woodruff keys may be used both on straight and tapered shafts.

The presence of a keyway in a shaft reduces its stiffness and especially its fatigue strength. The decrease in fatigue strength is much lessened when the internal corners of the keyway are given fillets. Even a small fillet radius gives an improvement. Keyway dimensions in gears, pulleys, couplings, etc., provided by manufacturers give adequate shear strength in the keys when these elements transmit their rated torque or horsepower at given revolutions per minute.

Other types of keys are the *gib head key* and the *pin key* (see Fig. 11.10).

The gib head key is a tapered square or flat key that has a protruding head to facilitate its removal. The head is a safety hazard, and in newer designs this type has been superseded by more modern approaches.

Pin keys may be cylindrical or tapered. They are easily assembled at the end of a shaft as a drive fit, and they are superior to square keys because they do not tip. The round groove has far better fatigue characteristics than the square keyway.

The optimal diameter for gib head keys and pin keys is approximately ¼ shaft diameter.

*Split taper bushings/internally tapered hub locking systems* (Fig. 11.11) are being increasingly used in conjunction with square keys to fasten any part provided with a hub. They are used not only for pulleys, sheaves, sprockets, and couplings but also for such parts as fan rotors, drums, and other similar shaft-mounted rotating parts for equipment and machinery.

They are clean-looking and easily removed. They can be reused indefinitely and leave no marks on the shaft.

There are several designs on the market under trade names such as QD and Taper-Lock. They all operate on the principle of drawing a some-

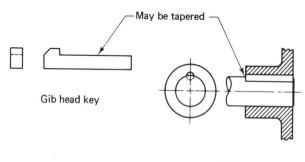

Gib head key

May be tapered

Pin key

**Figure 11.10**

what flexible, split taper bushing into a mating internal taper that is either machined directly into the hub or into a ring that may be welded or otherwise fastened to a wheel, sprocket, etc.

The split taper is forced in by the action of three pull-up screws. While entering the taper of the hub, the bushing is squeezed inward to press on and lock the shaft.

As shown in Fig. 11.11, the split taper bushing comes either with a flange (picture on the right) to accommodate the external hexagon head pull-up screws $A$ as well as the removal screws that enter hole $B$ when used or without a flange (picture on the left). In the latter case, the pull-up screws $C$ are headless and of the internal hexagon type. They are located in the common conical plane of the bushing and hub. Two other holes $E$ are provided to insert screws to remove the bushing when necessary.

The flangeless split taper bushing is somewhat more flexible and is also safer because the pull-up screws do not protrude beyond the face of the hub.

The operation of the flangeless split taper bushing is a little more difficult to understand than that of the flanged type. The headless pull-up screws $C$ are threaded in the hub but not in the bushing. When the pull-up screws rest on the end (bottom) of the slot in the bushing, additional rotation will push the bushing farther into the female taper to seat onto the shaft. To remove, the pull-up screws are removed, and two of them are inserted into the holes provided for removal $E$. These holes have thread in the bushing but not in the hub. In a way similar but opposite to the action of the pull-up screws, the bushing is now pulled out when the two screws are turned inward.

Removal screw holes should be kept filled with grease when the unit is assembled to exclude dirt.

To eliminate the weakening effect of the keyway on the shaft, and the expense of machining it, a European modification of the split taper bushing is used without a key and keyway. It is claimed to have adequate holding power while leaving no marks on the shaft.

Split taper bushings are economical because the same size tapered bushing can be made to fit a range of shaft sizes. This feature permits using a given hub for different diameter shafts, which reduces inventory requirements.

*Shear pins* are used whenever it is necessary to protect a transmission drive train from damaging overloads. The principle is to provide a purposely weak link somewhere in the system, which is under shear stress when the drive train is in operation. The link fails at a predetermined load. This load is much below that which would cause costly damage to the transmission elements such as gears, sprockets, chain, etc. After the pin has sheared off, the hub or flange in which it is located is permitted to idle

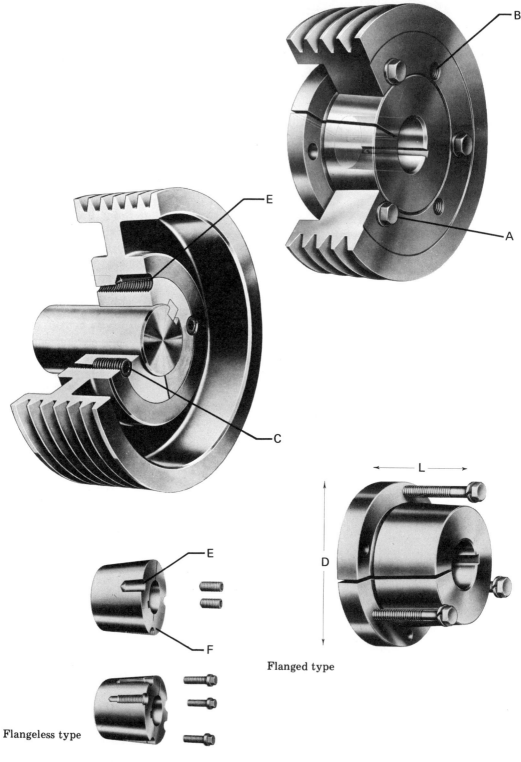

Flangeless type

Flanged type

**Figure 11.11**
The Gates Rubber Company, Denver, Colorado

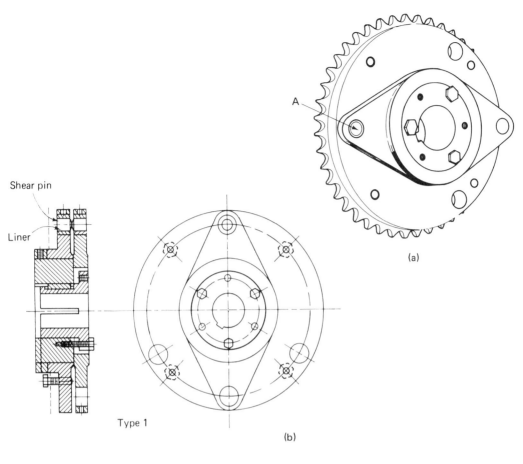

A

(a)

Shear pin

Liner

Type 1

(b)

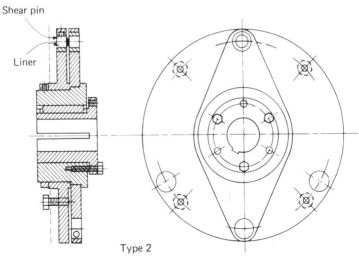

Shear pin

Liner

Type 2

**Figure 11.12**
Emerson Electric, Maysville, Kentucky

171

until the drive is shut down. A radial pin such as described earlier in this chapter (tapered or cylindrical) may be used as a shear pin, but a more common approach is to use shear pins in a position outside the hub, parallel to the axis of the shaft [see *A* in Fig. 11.12 (a)]. Here, a shear pin is used in a chain sprocket assembly, the hub of which is fastened to the shaft by a split taper bushing. The shear pin is undercut in the location of the desired shear plane.

After shearing of the pin, the inner part of the hub, which is keyed to the shaft, is torsionally disconnected from the outer part to which the sprocket is bolted. This permits the drive to idle. The assembly is grease-packed to assure low friction if the pin breaks.

## PROBLEMS

1. Why are opposite setscrews not recommended?
2. Explain why eccentric tapered lock pins do not weaken the shaft as much as radially placed pins of the same size.
3. Would you choose a spring pin or a groove pin for a sprocket assembly at the end of a shaft that must be removed at regular intervals?
4. What is the purpose of tapering square keys?
5. Can Woodruff keys be used as feather keys? Motivate your answer.
6. What are the advantages of pin keys? Would you recommend their use for applications that require frequent disassembly?
7. Describe two advantages of split taper bushing systems.
8. What are the two advantages of using a split taper bushing system that does not require a keyway?
9. What are the purpose and operating principle of shear pin assemblies?

# 12

# COUPLINGS, CLUTCHES, AND BRAKES

*Properties of Different Types of Couplings and Their Suitability to Various Service Conditions. Positive and Friction Type Clutches. Drum and Disc Brakes.*

Because of transportation and installation problems, the practical length of single shafts is limited. Longer shafts are usually obtained by means of *couplings* which are used to connect single lengths of shafting together. This assembly then acts as a single shaft. Couplings are also used to connect a shaft to a driver or to a driven unit or merely to connect a driver directly to its load, for example, to connect an electric motor to a centrifugal pump.

The following conditions may occur which influence the type of coupling selected:

*Radial or axial misalignment*, in which the axes of the shaft are parallel but do not coincide.

*Angular misalignment*, in which the axes of the shafts form a small angle with one another.

*Float*, in which there is a small amount of axial play between the shafts.

*Shock loads*, which are temporary, sudden, and transient torsional load peaks that depend on the nature of the driver and the driven equipment and/or the manner in which the loads are assumed by a transmission. Their severity is reflected by the magnitude of the *service factor* (see tables in Chapters 4 and 6).

*Torsional vibration*, which is a form of cyclical, repetitive shock load, such as typically produced by reciprocating machinery (IC engines, piston pumps, punch presses, etc.) The effect of torsional vibration is also incorporated in the service factor.

## 12.1 COUPLINGS

*RIGID COUPLINGS.* Rigid couplings provide solid connections between shafts, but they do not allow for either misalignment or float. This limits their application to cases in which proper alignment is lastingly ensured by close and firm support bearings.

Examples of *flange* and *sleeve* type rigid couplings are shown in Fig. 12.1 and Fig. 14.19. The flange coupling is available in a wide range of horsepower ratings and is easily disconnected by removing the shear bolts. Failure to maintain alignment of the shafts puts excessive loads on these bolts and may shorten their life.

The sleeve type is limited to shafts of equal diameter. It is easily removed in its entirety from the shafts.

Rigid couplings need no maintenance and have no wearing parts, but they do transmit all shock loads and torsional vibration from one shaft section to the other.

*FLEXIBLE COUPLINGS.* Couplings in this group usually permit one or both forms of misalignment and/or float. Most types also absorb shock and torsional vibration to a greater or lesser extent. All these

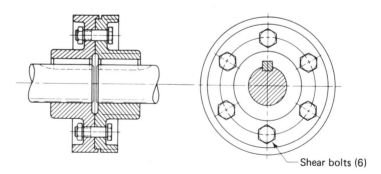

Shear bolts (6)

**Figure 12.1**

conditions produce internal and external friction in the couplings. This friction is transformed into heat to be dissipated to the surroundings.

*Flexible couplings with a nonmetallic flexing member* (usually made from rubber or some other elastomer) require neither lubrication nor maintenance but are mostly limited to low horsepower and revolutions per minute values. Their permissible temperature range is limited by the nature of the material of the flexing member. Examples are the *cushion* and *shear* types, which are discussed below.

The cushion type coupling (see Fig. 12.2) has its star-shaped flexing member loaded in compression. It permits some angular and axial misalignment, but it is only suitable for low horsepower applications.

Because of the greater possible displacement of the two coupling halves, which puts the flexing member into shear, the shear type flexible coupling (see Fig. 12.3) permits considerable misalignment (angular up to 4° and axial up to .13 in.) and float (up to .31 in.), depending on the size of the coupling and the duration of the shaft displacement (i.e., the time the misalignment condition lasts). For the same reason, the shear type coupling also absorbs shock loads and vibration well. It is suitable for loads ranging from fractional to approximately 600 hp.

In the catalog of one large manufacturer the selection of shear type couplings is on the basis of bore size and the tables give rated horsepower per 100 rpm for various service factors up to a maximum revolutions per minute value. This value decreases with increasing bore (i.e., coupling) size. For instance, a 1.125 bore coupling has a maximum of 4,500 rpm and a 9.000 bore coupling has a maximum of 810 rpm.

*Flexible couplings with a metallic flexing member* usually require lubrication. In most cases, grease is used. Due to centrifugal force, a protective cover is usually needed to prevent the lubricant from escaping.

Among the many types on the market, the following appear to be the most popular: chain coupling, gear coupling, Falk Steelflex coupling, fluid (hydraulic) coupling, and universal joints. These are discussed below.

The *chain coupling* (single or double roller types) (see Fig. 12.4) consists essentially of two identical sprockets that are held together by a length of chain that has as many rollers as there are teeth on the sprockets.

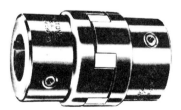

**Figure 12.2**
Boston Gear, Quincy, Mass.

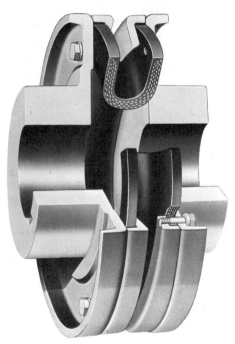

Falk Corporation, Milwaukee, Wisconsin

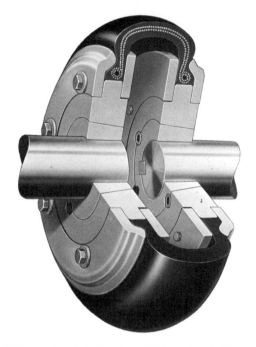

Reliance Electric Co., Inc., Mishawaka, Indiana

Figure 12.3

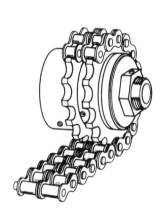

Figure 12.4
Reliance Electric Co., Inc., Mishawaka, Indiana

Chain coupling accepts limited angular misalignment ($\frac{1}{2}°$ to $1\frac{1}{2}°$), axial (from .010 in. to 2% of pitch), and float (.020 in. to .070 in.), depending on coupling size.

In similarity with chain transmissions, chain couplings may be fairly noisy and absorb neither shock loads nor torsional vibration. Their primary advantage lies in the fact that chain removal for disconnection or chain replacement can be effected merely by pulling one pin so that shafts and sprockets need not be disturbed.

Since chain and sprockets (in shear and tension) are much stronger than the shaft (in torsion), selection of chain couplings is often done on the basis of maximum bore (shaft diameter).

The *gear coupling* (see Fig. 12.5) is somewhat similar in construction to the chain coupling. Two identical pinions are encased in a mating internal gear of the same tooth number and pitch. The face and flank of the pinion teeth are given a certain curvature in the direction of the face width (see Fig. 5.4), called *crown*, which permits some angular and axial misalignment.

Gear couplings are compact, silent in operation, and have a wide load range.

The misalignment, float, and damping characteristics of gear couplings are much like those for chain couplings, and they are selected on the same basis.

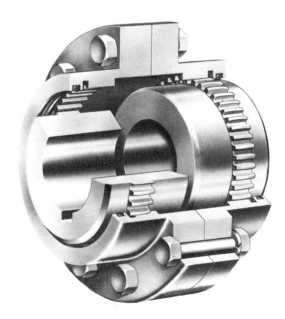

**Figure 12.5**
Ajax Flexible Coupling Co., Westfield, New York

The *Falk Steelflex coupling* (see Fig. 12.6) uses a continuous spring steel strip that weaves back and forth in serrations cut in the periphery (i.e., the circumference) of the coupling halves' flanges. The serrations are tapered out somewhat toward one another, which permits some small relative movement between the two halves, thus providing excellent shock and vibration absorption.

The Falk coupling allows approximately 1° of angular misalignment, ½% of shaft diameter radial misalignment, and from .125 in. to .25 in. float. It has a very wide load range and may be operated at speeds up to 6,000 rpm, depending on size.

The *fluid (hydraulic) coupling* (see Figs. 12.7 and 12.8) is in a class by itself. It uses a fluid for the transmission of torque. The input half of the coupling, by means of its radial vanes, imparts kinetic energy to the fluid, which is absorbed by the output half, except for a small portion that is lost to shear and friction of the fluid. This portion is converted into

**Figure 12.6**
Falk Corporation, Milwaukee, Wisconsin

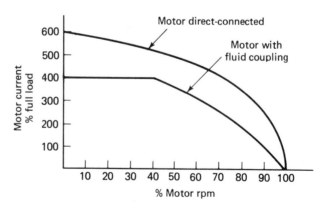

**Figure 12.7**

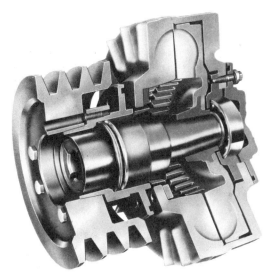

Figure 12.8
Dana Corporation, Warren, Michigan

heat and is dissipated to the surrounding air via the coupling's housing, which has cooling ribs for that purpose.

The halves are only coupled when the driver is rotating, and to transmit torque at all there must be some slip between the two halves, typically from 3% to 6%. Maximum torque occurs at stall of the output half.

Since power output of the driver rotor varies with the *cube* of the revolutions per minute, manufacturer's ratings are tied to a revolutions per minute range from which no deviations should be made. In the coupling illustrated, the torque characteristics may be altered by varying the fluid level. By their nature, fluid couplings cannot absorb any misalignment and permit very little float, but they do absorb shock and torsional vibration to a high degree.

Because they stall on reaching maximum torque, fluid couplings give excellent overload protection, although forced cooling must be provided if prolonged overload periods are to be expected. With the addition of a stator between the two coupling halves, a fluid coupling becomes a torque converter and may be used as a variable speed drive. A typical example of the use of such a unit is found in certain types of automotive transmissions (see Chapter 8). Fluid couplings have the characteristic of very smooth load assumption when they are used to connect an electric motor to a slowly starting load, and thus reduce peak loads on the electrical system as well as on the transmission (see curve of Fig. 12.7). They also provide an easy means of balancing the load when more than one prime mover is to drive a transmission system.

Lastly, a fluid coupling may be easily and quickly disconnected from the system merely by draining the fluid.

   The *universal joint* is a coupling especially designed for considerable (intentional) angular misalignment. The commonly used Hooke or Cardan joint (for instance, in automobile propeller shafts), shown in Fig. 12.9, has the drawback that it does not transmit an entirely constant angular velocity (or revolutions per minute). The effects of this characteristic may be reduced by putting two Hooke joints in series, with the jaws on the intermediate shaft lying in one plane (see Fig. 12.10). In this manner, the acceleration originating in the first joint is balanced by the simultaneous deceleration in the second one. For optimum effect, angle $A$ should be as nearly equal to angle $B$ as possible.

   Newer types of universal joint, such as the Rzeppa and Weiss couplings (see Fig. 12.11), have been developed which do not possess the drawback described above. In all these designs the contact between the coupling driver and follower is made in a plane that bisects the angle the two shafts form with one another (the *homokinetic plane*) (see Fig. 12.12). This condition ensures constant velocity transmission.

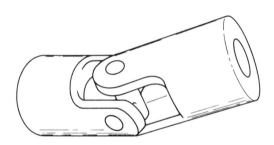

**Figure 12.9**
Curtis Universal Joint Co., Inc., Springfield, Massachusetts

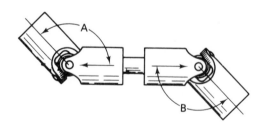

Correct installation of
two universal joints on common shaft

**Figure 12.10**

(a)

(b)

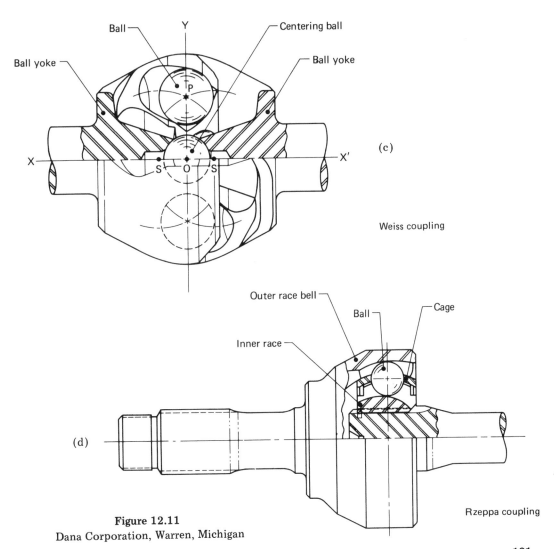

(c)

Weiss coupling

(d)

Rzeppa coupling

**Figure 12.11**
Dana Corporation, Warren, Michigan

181

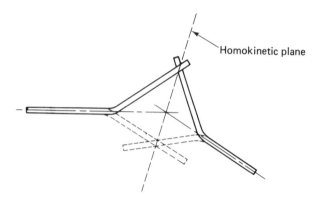

An elementary but kinematically perfect
universal joint.

**Figure 12.12**

## 12.2 CLUTCHES

Clutches permit the connection (disconnection) of a torsional load with (from) a driver when either side of the clutch is either rotating or stationary. Clutches may thus transmit load in both directions, except as noted. Since clutches are designed to transmit a given torque, horsepower ratings must be accompanied by a certain revolutions per minute value, for instance, "80 hp at 500 rpm" (see Chapter 4).

*Positive clutches* do not slip, are small and inexpensive, and do not generate heat, but they cause shock loads that may be severe when engaged with one part rotating and the other stationary. For this reason, using positive clutches for running engagement is mostly limited to applications involving small loads and low revolutions per minute. Two types of positive clutch are illustrated in Fig. 12.13. The clutch marked (b) is technically of the overrunning type (see below) and transmits torque in one sense (i.e., direction of rotation) only.

*Friction clutches* permit gradual assumption of the load and are used when stopping the driver is impossible or impractical. This is for instance the case when the prime mover is an IC engine. Friction clutches may also be used as an overload protective device by designing them so that they start to slip when a predetermined torque has been reached and by slipping momentarily whenever a shock load occurs. Due to the gradual load assumption characteristic of these clutches, much higher driver revolutions per minute assumptions are possible than with positive clutches.

The main categories of friction clutch are the *disc* and the *rim* types. The friction member of the disc type consists of one or more friction

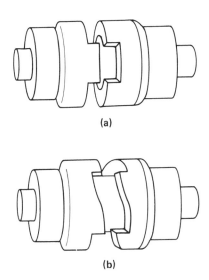

(a)

(b)

**Figure 12.13**

discs which move axially during engagement. The single disc type is usually used dry, and the disc is covered with a high-friction compound. The familiar clutch of manual-shift passenger cars is of this type.

Multiple-disc friction clutches often run in oil. The oil provides smoother engagement and easier disengagement and also provides a means of removing frictional heat. Discs in multiple-disc clutches running in oil are mostly uncovered and are made of heat-treated steel. The type is used in heavy-duty applications such as truck transmissions.

Disc-type clutches have the advantage of large friction surfaces for a given size and weight and uniform pressure distribution over the contact area of the disc(s).

An example of the rim (or drum) type is the *Fawick Airflex clutch* shown in Figs. 12.14. and 12.15. This type, which is widely used in industry, consists essentially of two short, concentric cylinders ("rims") whose diameters are sufficiently different to permit a flattened, annular tube of rubberized canvas to be located between them. This annular tube, which is fastened to the outer rim, can be made to expand under air pressure; thus it presses on the inner rim. In this manner, the tube makes frictional contact between the outer (driver) rim and the inner (driven) rim.

The air pressure, which must be sufficiently high to overcome the effects of centrifugal force on the tube, may be modulated to vary the engagement characteristics and also to limit the torsion value at which slipping (for overload protection) begins. When the air pressure is released from the tube, disengagement of the latter is assisted by centrifugal

**Figure 12.14**
Eaton Corporation, Cleveland, Ohio

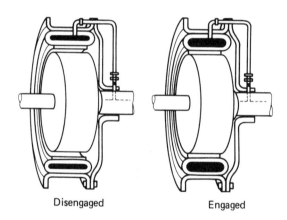

Disengaged                    Engaged

**Figure 12.15**
Eaton Corporation, Cleveland, Ohio

force when rotating. This type of friction clutch automatically compensates for wear.

*Overrunning clutches* provide torque transmission in one sense only. An elementary type, suitable for relatively low torque and revolutions per minute values only, is the *ratchet clutch*, which, among others, is used in the well-known free-wheeling driving sprocket of bicycles. The *roller type* clutch (Fig. 12.16) is often used in industrial equipment. Under the influence of torque and rotation of the driver shaft, a number of rollers

184

**Figure 12.16**
The Torrington Company, Torrington, Connecticut

held in a cylindrical cage around the shaft may be wedged between it and the cam-shaped, mainly concentric recesses inside the clutch body. While the rollers are wedged, the clutch acts as if it were a rigid coupling.

In another, somewhat similar type, the camming action is provided by the cam-shaped rollers themselves (here called *sprags*) (see Fig. 12.17). In both types the camming action is possible in one sense only, so that the driven part may run faster than the driver (hence "overrunning"), either by accelerating the driven part or by slowing down or stopping the driver shaft.

*Centrifugal clutches* engage the load only after a certain driver revolutions per minute have been reached. They are useful in equipment that has high inertia load (i.e., that have heavy rotational masses that are slow to accelerate) because they permit AC induction motors (the cheapest and most generally used motors in industry) to assume their load at a point of much higher torque than available at standstill. As in the case of fluid couplings, the motors can thus often be smaller than needed when directly connected to their loads, which in turn results in lower peak loads on the electrical system. These clutches also provide automatic disconnection of an idle driver unit from a multiple prime mover installation. When an overload causes the prime mover to slow down, centrifugal force is reduced, and the clutch starts to slip, thus giving overload protection.

Prolonged overload periods may require forced cooling. Several mechanical centrifugal clutches employ levers and/or weights acting against a drum to provide clutching action. The Flexidine ® coupling in Fig. 12.18 uses heat-treated shot, which acts as a dry fluid, as the torque-transmitting medium. Centrifugal force throws the charge to the

**Figure 12.17**
Dana Corporation, Warren, Michigan

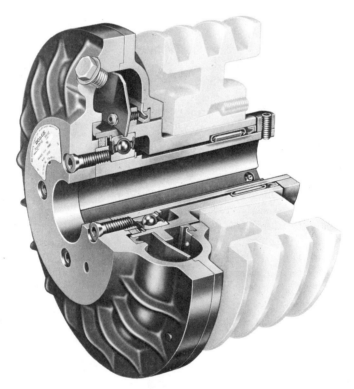

**Figure 12.18**
Reliance Electric Co., Inc., Mishawaka, Indiana

perimeter of the housing and packs it densely around the articulated (scalloped) rim of the follower rotor, which in this manner is entrained. The maximum torque (beyond which the coupling will start to slip and provide overload protection) may be modulated by varying the quantity of the charge.

## 12.3 BRAKES

Brakes are used for the gradual slowing down and stopping of rotating masses, and in some cases, for the prevention of rotation ("holding brakes").

Brakes are often very similar to clutches in construction, which should not be surprising since in essence a brake is a clutch with one side held stationary.

The rotational energy absorbed is transformed by friction into heat that must be dissipated, usually by convection and radiation to the surroundings. As with clutches, repeated start and stop cycles may require some form of cooling system.

*MECHANICAL BRAKES.* Mechanical brakes use friction to accomplish their purpose. The lining material that provides the friction may be organic (most common), metallic, or ceramic. The materials are used either separately or in combination. An often-used friction material consists mainly of asbestos. Since asbestos fibers in the atmosphere, worn from linings, are now recognized as a serious long-range health hazard, it may be expected that asbestos compounds will eventually be replaced entirely by other materials such as metallic and ceramic linings, which at present are used primarily for extreme-duty conditions. Linings may be bonded or riveted in place. Bonding provides greater *available* thickness for wear, but it involves a somewhat more demanding and expensive manufacturing process.

Actuation of mechanical brakes may be by mechanical, hydraulic, pneumatic, or electrical means, again in similarity with clutches. One of the oldest mechanical brakes is the *band brake* (see Fig. 12.19). It

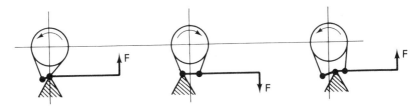

**Figure 12.19**

can be made self-energizing (i.e., it can generate part of its actuating force) for a given sense of rotation by a certain arrangement of the fulcrum points. Today the band brake is mainly used as a holding brake.

*Drum brakes* have external or internal brake shoes acting on the drum. Internal brake shoes are used, among others, in automobiles. As with the band brake, the drum brake can be made self-energizing for a given sense of rotation and in a similar manner (location of fulcrum points).

Two types of *disc brake* are the *plate type*, which is similar in design and operation to a single-disc friction clutch, and the *single-disc caliper type* (see Fig. 12.20). In the single-disc caliper type brake, two opposed

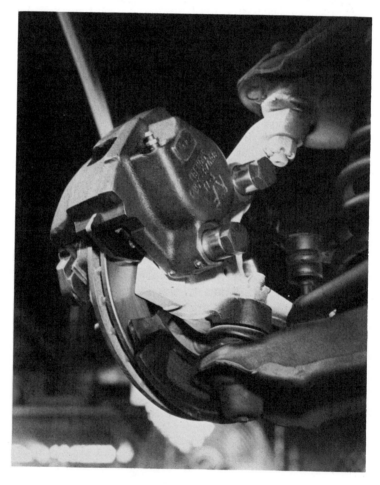

**Figure 12.20**
Bendix Corporation, South Bend, Indiana

friction pads are made to press simultaneously on opposite sides of a disc. It is by far the most commonly used type. Originally designed for aircraft wheels, it is rapidly displacing the drum type in automobiles because it provides the following advantages: simple design, little sensitivity to contamination by water, oil, or grease, resistance to fading, and superior heat dissipation ability.

Because of the relatively small pad area, the actuation force is so high that some form of external boosting is usually necessary. In its basic design, the single-disc caliper type brake is not self-energizing, however, the Chrysler Corporation has developed a self-energizing design.

*ELECTRIC BRAKES. True* electric brakes use electromagnetic forces to provide a braking effect. Electrically *actuated* brakes use these same forces to engage friction (mechanical) brakes. Examples of true electric brakes are eddy current and magnetic particle brakes. They are described below.

The *eddy current brake* has a steel disc which, upon actuation of the brake, is subjected to a strong magnetic field. The rotating disc generates strong eddy currents in its surface, which in turn create a magnetic field opposing the disc's rotation. Since there is no frictional contact, the eddy current brake does not wear, but it does need an additional holding brake because some rotation is needed to generate a braking effect.

The *magnetic particle brake* uses magnetized ferrous particles either in a slurry or dry to provide braking and holding power between a ferromagnetic housing and disc. Friction occurs during periods of disengagement. In some cases, this may be a drawback.

## PROBLEMS

1. On size C Vellum make a full-scale layout of the small conveyor drive schematically shown in Fig. 12.21. Show front and top views. Center the drawing as well as you can.

   (*a*) Boston ratiomotor, F315, 50 ratio. 1/4 hp, 115V AC, output rpm: 35; torque: 248 in.-lb. Dimensions as per sketches in Fig. 12.22.

   (*b*) Universal joints, UJNS 10-10, hole diameter: .625 in. nominal; overall length: 5 in.; hub diameter 1.5 in.; hub projections; .875 in.; keyway: .188 × .093 in. *Note:* Since the output shaft of the ratiomotor is .500 in. nominal diameter, a bushing is needed to bring it up to .625 in. nominal diameter. See the sketch of the universal joint in Fig. 12.23.

   (*c*) .625 nominal diameter steel shaft to fit.

   (*d*) As C, except show short length only.

2. What conditions must be considered when a coupling is being selected?

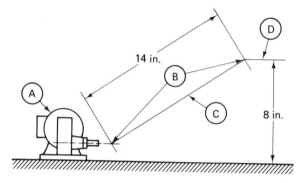

**Figure 12.21**

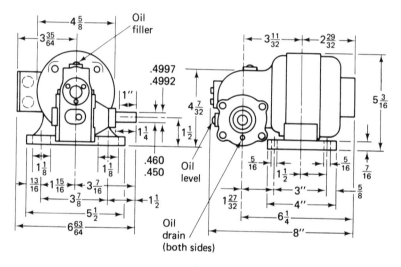

**Figure 12.22**
Boston Gear, Quincy, Mass.

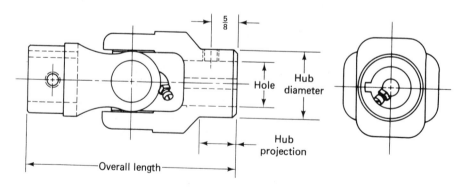

**Figure 12.23**
Boston Gear, Quincy, Mass.

3. Why are chain and gear type couplings rated by their bore (i.e., shaft size)?

4. Why can a fluid coupling never be used for a synchronous drive?

5. What is the drawback of the Hooke joint and how can it be alleviated?

6. Can overrunning clutches be used for overload protection? Motivate your answer.

7. A clutch has a rating of 35 hp at 700 rpm. Compute its rating at 575 rpm.

8. What are the advantages of single-disc caliper type brakes?

9. What qualification must be made when "electrical brakes" are specified?

# 13

# CAMS

*Displacement Diagrams. Point and Roller Followers. Constant Velocity, Harmonic Motion, and Constant Acceleration and Deceleration Cams. Positive Motion Cams. Plate and Cylindrical Cams. Combined Motion Obtained from Two Synchronized Plate Cams.*

A cam and its follower form a mechanism that converts rotary motion (usually constant speed) or oscillating motion into a cyclical (repetitive) linear or angular motion.

*Plate cams* are made in the shape of a flat plate with either a contoured periphery or a circular one that has a contoured groove in one face, and a hub. Plate cams rotate around an axis that is perpendicular to the plane of the plate and they impart mainly radial motion to their followers (see Fig. 13.1).

*Cylindrical cams* (Fig. 13.2) consist of a cylindrical body designed to rotate around its axis. Either one end or a peripheral groove is shaped in such a manner as to impart the desired motion to a follower in a direction mainly parallel to the cylinder's axis.

There are other types of cam, but they are outside of the scope of this chapter.

Both plate and cylindrical cams have a wide field of application, and they belong to the most efficient mechanisms known. For example, they

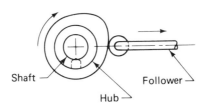

Shaft

Follower

Hub

Figure 13.1

Figure 13.2

are used in automatic screw machines, packaging machinery, coil and bobbin winders, tool and die design, and many indexing devices, such as the one shown in Fig. 13.3.

Cams are indispensable for the valve actuation in both gasoline and diesel IC engines. Two plate cams acting on a single follower, one in the

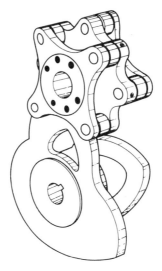

Figure 13.3
Amcam Corporation, Farmington, Connecticut

*x*-direction and the other in the *y*-direction, can produce almost any type of path (see Chapter 2).

Plate cam *followers* may be *radial*, in which case the centerline of the follower intersects with the axis of rotation of the cam, *offset* when this is not the case, and *swinging* when the follower is hinged around an axis outside of the cam's surface so that it describes a circular arc (see Fig. 13.4).

The point of contact between cam and follower may be in the shape of a point, a roller, or a flat or rounded surface, as shown in Fig. 13.5.

To design a plate cam, we must first establish the motion requirements. Most of the time these are recorded in the form of a *displacement diagram* but sometimes in the form of a displacement *schedule* or *table*. Such a diagram shows the desired displacements of the follower for a number of equal, sequential angular displacements of the cam, which are usually expressed in degrees.

In addition, we must establish the radius of the *base circle*, which is numerically equal to the center distance of the follower contact point around the periphery nearest to the center of rotation of the cam. The base circle is the basis of the cam outline, to which are added the sequential distances the point of contact of the follower is required to travel, as indicated by the displacement diagram. The diameter of the base circle must be at least equal to the diameter of the hub. In the case of cams made integral with the shaft, such as is the case in IC engine camshafts, the base circle diameter must be at least equal to the shaft diameter.

In the example shown in Fig. 13.6, it is required that a pointed, radially translating follower (i.e., one moving in a straight line away from the center of the cam) move .75 in. at constant velocity for the first 120° of counterclockwise cam rotation. The follower must then stay in that position ("dwell") for 30° and then return to its starting position, again at constant velocity. That position must be reached at 210°. The follower should then dwell until the starting point at 360° (= 0°) is reached. With this information, and using 15° increments of cam rotation, we can draw

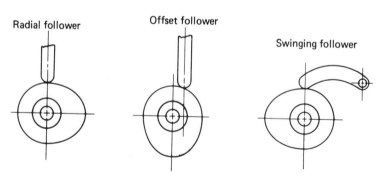

Radial follower    Offset follower    Swinging follower

**Figure 13.4**

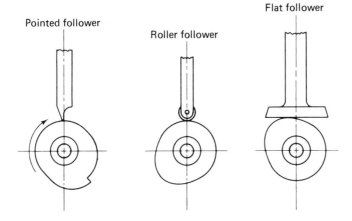

Pointed follower

Roller follower

Flat follower

Figure 13.5

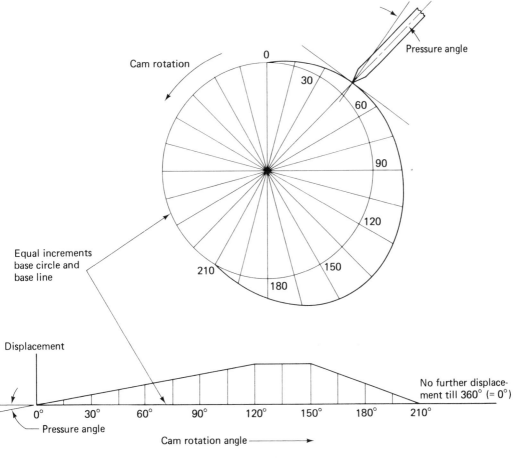

Figure 13.6

the displacement diagram shown in Fig. 13.6 at a reduced scale. After completing the displacement diagram, we can construct the cam proper after the radius of the base circle has been determined. In this example a base circle radius of 2 in. is used.

To draw the cam, begin with drawing the base circle. Allow enough space around it for the cam's outline. Using one 30-60-90 and one 45-45-90 triangle in the manner shown in Fig. 13.7, or a protractor, accurately

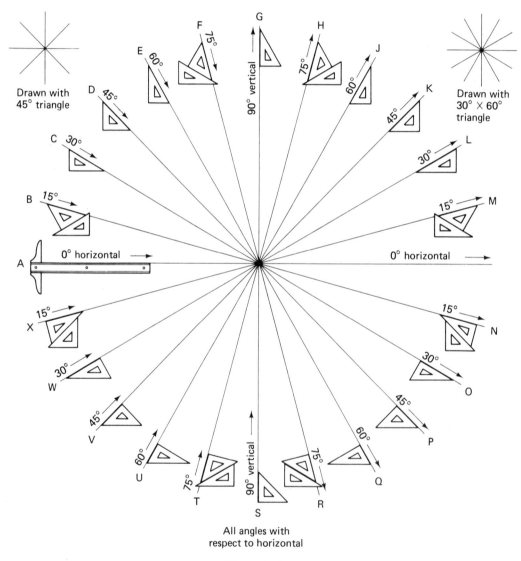

All angles with
respect to horizontal

**Figure 13.7**
Macmillan Publishing Co., Inc., New York, New York

divide the circle into $15°$-sectors as shown. Draw radii at each multiple of $15°$, extending beyond the base circle as required. With the dividers, transfer each individual displacement distance from each multiple of $15°$ on the displacement diagram to its corresponding radius. Measure outward from the base circle. Note that the construction proceeds in a clockwise direction, i.e., against the specified sense of rotation of the cam. After all distances have been transferred, draw a smooth line through all the displacement points with the aid of a French curve. The follower will now be in the required position at any point of cam rotation.

The angle between the normal to the cam at any point, and the radius at that point, is called the *pressure angle*. It is shown in Fig. 13.8. Note that the pressure angle also occurs in the displacement diagram and will be shown on the right scale if the length of the divisions of the displacement

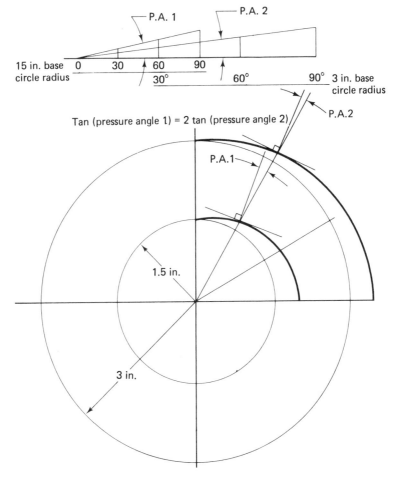

**Figure 13.8**

diagram is equal to that of the divisions of the base circle. In the example, a base circle of 2 in. radius would require divisions of a length equal to

$$\frac{15°}{360°} \times 2\pi r = .523 \text{ in.}$$

Increasing the base circle radius reduces the pressure angle for a given displacement of the follower, as shown in Fig. 13.8. It will be seen in the displacement diagrams in Fig. 13.8 that

$$\text{tan (pressure angle)} = \frac{\text{displacement}}{\text{length of arc of base circle subtending total rise}}$$

from which the pressure angle can be found.

To prevent the possibility of binding for a pointed follower, the pressure angle is usually not permitted to exceed 30°.

The construction of a cam for a *roller-tipped*, translating follower is slightly different. The path of the *center* of the roller is that indicated by the displacement diagram, and this curve, called the *pitch line*, can be drawn by using the method similar to that followed for a pointed follower, except that the base circle must now have a radius larger by one-half the roller diameter.

To find the actual outline of the cam, draw a number of circles having a diameter equal to that of the roller with their center on the pitch line. Then construct the cam contour by drawing the inside curve tangent to all these circles. The number of circles depends on the rate of change in the pitch curve at any point: the higher the rate, the greater the number of circles. When there is no change, no circles are needed, since the cam outline there will be a circular arc. Figure 13.9 illustrates this procedure. It is for a cam similar to the one shown in Fig. 13.6, but it has a roller follower instead of a pointed one.

An *offset* follower cam construction is shown in Fig. 13.10. A roller-tipped follower is used, which is the type most frequently employed in this case. The displacement diagram used is again that of Fig. 13.6. To the layout of the base circle (its radius including the roller radius) is added the centerline of the follower in such a manner that the point of intersection of the centerline with the base circle is assigned 0°, the starting point of construction. The required follower displacements at desired intervals (here 30° for clarity) are now transferred from the displacement diagram to the radius at 0°. Concentric arcs of appropriate length are now drawn through these points as shown and then pitch point offsets *a*, *b*, *c*, etc., are marked off on them. The location of the roller centers are thus established, so that the pitch line can be drawn. As many circles representing the roller outline as needed for accurate cam construction can now be drawn and subsequently we can draw the cam itself.

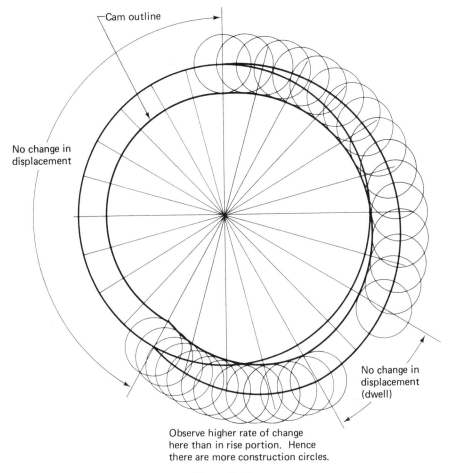

Cam outline

No change in
displacement

No change in
displacement
(dwell)

Observe higher rate of change
here than in rise portion. Hence
there are more construction circles.

Figure 13.9

A *swinging roller* follower cam is constructed in a similar manner, but here the arc of swing of the center of the roller is used instead of the centerline of the follower.

The construction of a cam with a *flat-faced*, translating follower is shown in Fig. 13.11. The displacement diagram is that of Fig. 13.6. Note that since the point of contact with the cam may be anywhere on the flat face of the follower, the construction of a radial and an offset, flat-faced, translating follower cam is the same.

Let us at this point compare the action of pointed, roller-tipped, and flat-faced radial followers. Consider a cam as shown in Fig. 13.12 with an abrupt change in displacement. Note that only the specially shaped,

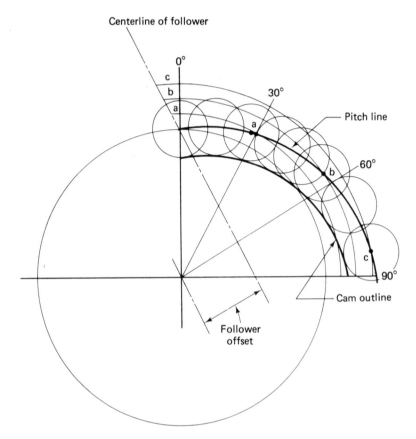

**Figure 13.10**

pointed follower shown is able to follow the cam outline exactly. A roller-tipped follower will describe an arc when rounding the sharp tip of the cam, while a flat follower will not be able to follow the concave portion of the cam at all. We may thus conclude from the foregoing that the pointed-tip follower is the most sensitive and that the flat-faced follower is the least sensitive to changes in cam outline. The flat-faced follower is able to follow very simple cam outlines only.

The sudden velocity changes imparted to the follower by the constant velocity cam thus far discussed do not make this type very well suited to high revolutions per minute requirements.

Several kinds of curve providing follower motion that is smoother than that of the constant velocity type are in existence, but only a few can be mathematically defined. We shall discuss two examples of the latter group which, because of their easy construction, are in general use. They are the harmonic motion curve and the constant acceleration and deceleration curve.

200

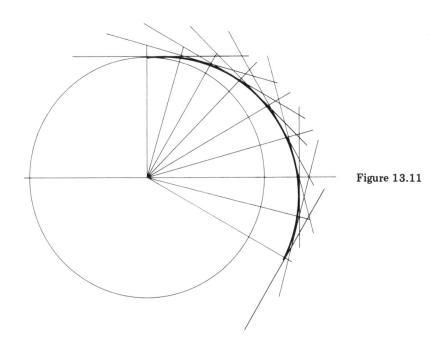

Figure 13.11

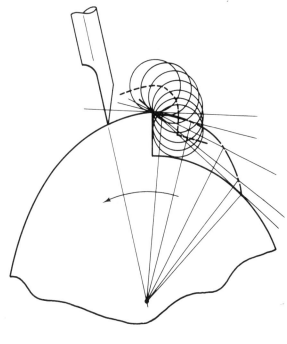

Figure 13.12

---------------- Path of roller-tipped follower

--·---·---·--- Path of flat-faced follower

──────────── Cam outline-path of pointed follower

201

In *harmonic motion* the shape of the rise and fall portions of the displacement diagram is that of the curve defining the sine function (y = sine *x*), hence it is also called sinusoidal. The motion is similar to that of the Scotch yoke (see Chapter 2). The construction of the harmonic motion curve in the displacement diagram is shown in Fig. 13.13. Note that the 30°-increments of the construction circle do not bear any relation to the increments of the displacement diagram. The latter should be one-sixth of the total angular cam displacement during rise or fall. Also observe that the total rise or fall of the cam is equal to the diameter of the construction circle.

It will be obvious from inspection of the harmonic motion curve as compared with the constant velocity curve (superimposed on the drawing as a dotted line) that much smoother motion is imparted to the follower by the former, for velocity changes at 0° and 120° are gradual rather than sudden.

In *constant acceleration and deceleration motion* the distances the follower travels during rise or fall are proportional to those given by the equation $S = \frac{1}{2}at^2$, in which *a* is the acceleration and *t* is the time in seconds or other unit of time. Since this equation mathematically defines the parabola, this motion is also called *parabolic*. Assume that the entire motion takes 6 s. During the first half of the motion (3 s), *a* is an *acceleration* so that the velocity of the follower *increases*. Since $\frac{1}{2}a$ is constant, distance S is proportional to (not equal to) $1^2 = 1$, $2^2 = 4$, and $3^2 = 9$. Successive increments in follower travel (*S*) are then proportional to

$$S_1 - S_0 = 1 - 0 = 1$$

$$S_2 - S_1 = 4 - 1 = 3$$

$$S_3 - S_2 = 9 - 4 = 5$$

In the second half of the motion period the action is reversed: *a* becomes a *deceleration* and the follower velocity *decreases*.

The increments of *S* also decrease in reversed order, so that

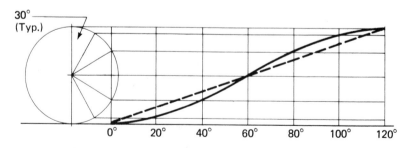

**Figure 13.13**

$$S_4 - S_3 = 5$$

$$S_5 - S_4 = 3$$

$$S_6 - S_5 = 1$$

As in the construction for harmonic motion, we thus again have six equal angular increments providing points for the displacement curve [at constant revolutions per minute of the cam, equal time increments mean equal angular (rotational) increments].

To construct the curve on the displacement diagram, we draw these incremental lengths (1, 3, 5, 5, 3, 1) on any convenient scale on a line at approximately $315°$ with the $y$-axis, as shown. The total required displacement of the follower is then marked off on the y-axis and a line is drawn connecting this mark with the highest mark on the $315°$ line.

Lines parallel to this line through the various incremental points and intersecting with the y-axis provide the division of the displacement distance into sections with the same ratio. Construction of the displacement curve from then on is similar to that for the harmonic curve.

The motion obtained by the constant acceleration and deceleration curve is not as smooth as that provided by harmonic motion.

With ordinary plate cams, the outward (rising) motion of the follower is positive, i.e., it is pushed outward directly by the cam. However, the return motion must be externally energized. In the most elementary, slow-acting systems the follower can often return by gravity alone. In most cases, however, the energy for the return motion is supplied by a spring, which obtains this energy during the outward movement of the follower. Essentially then, the total force exerted by the cam on the follower during the outward motion is that needed for the work performed by the follower during this motion *plus* that needed for the follower's return. In the case of high-velocity cam action, the return force needed may be very high.

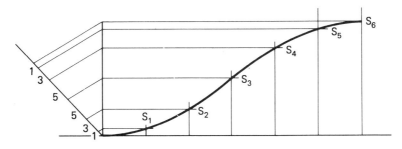

Figure 13.14

Since there exists mainly line contact between cam and follower tip (flat or roller; pointed tips would not be considered here), very high, transient local stresses may thus be induced at the contact point during rise of the cam, which may lead to early fatigue failure of the active surfaces of both cam and follower. For this reason, other designs of cams have been developed for heavy loads in which both outward and return motion are directly caused by cam action. Such cams are said to have *constrained motion.*

An example of a constrained motion type is the *grooved plate cam.* This cam has a circular plate with a milled groove of suitable depth and of a width slightly larger than the roller follower diameter (see Fig. 13.15). The path of the roller is that of the centerline or pitch line of the groove, which is also that of the end mill cutting the groove.

There is some lost motion (backlash) in this system because the groove must be slightly wider than the roller diameter. If this cannot be tolerated, two rollers of slightly different diameter running in a stepped groove may be used (Fig. 13.16).

Another positive motion system is the pair of *matched plate cams,* each with its own roller mounted on a common translating follower. Matched plate cams can be made to produce minimum backlash. The plate cam indexing mechanism shown in Fig. 13.3 is a special form of matched plate cams.

*Cylindrical cams* that have a peripheral groove in which a follower tip is engaged (Fig. 13.2) are another type of constrained motion cam. The displacement diagram can be considered as a schematic representation of the flattened cylindrical outside, the displacement line being the pitch line of the groove. The outline of the groove can be found by drawing many circles of a diameter equal to the follower diameter and astride the pitch line (Fig. 13.17). The groove outline is formed by the two curves tangent to the circles.

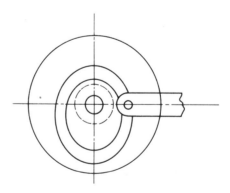

**Figure 13.15**

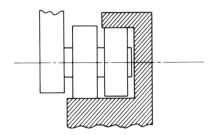

Figure 13.16

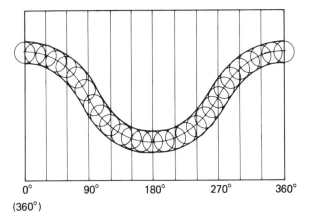

0°       90°      180°      270°      360°
(360°)

Figure 13.17

Grooved cylindrical cams have the same drawback as the grooved plate cams mentioned earlier.

*Manufacture of cams.* The most accurate and advanced method of cam manufacture, used for the production of master cams and in cases where extreme accuracy is essential, is by continuous-path numerical control of precision milling and grinding machines. Cams of lower accuracy may be made on contouring type milling machines which use a master cam as a template. Quality cams may be made from high-chrome, air-hardening tool steel, a material that has great resistance to wear.

## PROBLEMS

1. Construct a constant velocity cam for clockwise rotation with a .75 diameter radial roller follower. (Remember that the construction of a clockwise cam proceeds in a counterclockwise direction, and vice versa). The follower is to rise 1 in. at constant velocity over the first

$90°$ of cam rotation, then dwell for $15°$, and then return to its start-ing position after $180°$ of cam rotation. It should then dwell until the starting point $(360° = 0°)$ is reached again. The base circle dia-meter is $= 6$ in. First draw the displacement diagram $(15° = .5$ in.$)$, then the construction circle $(R = 3.375$ in.$)$, then the pitch line, then the circles representing the roller positions, and finally the cam outline.

2. Design a cam for counterclockwise operation for use with a 1 in.-diameter radial roller follower. The base circle diameter is 6 in. The follower is to rise 1.5 in. over $90°$ of cam rotation and return to its starting position at $180°$. The follower must then rise again for a dis-tance of 1 in. reached at $270°$ and then return to its starting position at $360° = 0°$. All motions are to be harmonic. Remember to draw a 7-in. diameter construction circle so as to end up with a 6-in. dia-meter base circle.

3. Design a constant acceleration and deceleration cam with a 6-in. base circle for use with a 1-in. diameter radial roller follower, a 1-in. dia-meter shaft, and a 2.5-in. diameter hub. The hub is keyed to the shaft with a .188-in. square key. The cam is to rotate counterclock-wise. The follower is to rise 2 in. over $120°$ of cam rotation in har-monic motion, then dwell for $60°$, then return to its starting position at $300°$ with constant acceleration and deceleration motion, and finally dwell to $360° = 0°$.

4. On size B paper draw a baseline ($x$-axis) approximately 8 in. long and mark off six parts, each one 1¼ in. long. Erect a 4½-in. perpendicular line ($y$-axis) 4½ in. long at its left end. Draw a half circle ($R = 2$ in.$)$ tangent to the $x$-axis, with its center on the $y$-axis, and pointing to the left. Draw radians at $30°$, $60°$, $90°$, $120°$, and $150°$ and construct, as accurately as possible, a harmonic motion curve (see Fig. 13.13). Now draw a line approximately 5 in. long through the intersection of the $x$- and $y$-axes pointing upward and to the left at about $45°$ to the $y$-axis. Starting at the intersection, divide this line into sections .25 in., .75 in., 1.25 in., 1.25 in., .75 in., and .25 in. long. Following the method demonstrated in Fig. 13.14, accurately construct a constant acceleration and deceleration curve. Which of these two motion curves provides the smoother motion?

5. A part in a packaging machine has to move in one direction for 2 in., then move sideways for 1.5 in., and then retrace its path. The moving part is to make at least 120 cycles per minute. It is proposed to use two synchronized plate cams turning counterclockwise to obtain this motion sequence. The cams are to act upon two 1-in. roller-tipped, springloaded slides, one on top of the other and moving at right angles. The general arrangement is to be as shown in Fig. 13.18. $A$ and $B$ are cam centers. The base circle diameter is to be 5 in., the

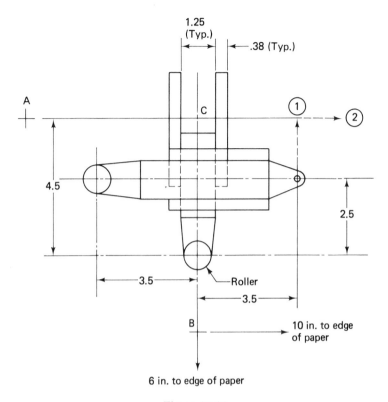

1.25
(Typ.)

.38 (Typ.)

A

C

1

2

4.5

2.5

3.5

Roller

3.5

B

10 in. to edge
of paper

6 in. to edge of paper

**Figure 13.18**

shaft diameter is to be 1 in., and the hub diameter is to be 2 in. Use harmonic motion and design the cams. Use size C vellum.

6. What is the hazard in having too great a pressure angle when using a pointed follower? What can be done to reduce this angle?

7. What is the purpose of constrained action cams?

8. What is the difference between the action of a grooved plate cam and the action of a stepped groove cam?

# 14

# OCCUPATIONAL SAFETY AND HEALTH ACT

*Some OSHA Requirements Pertaining to Mechanical Power Transmission Elements.*

In 1970 the Williams-Steiger Occupational Safety and Health Act was passed by the federal government "to assure so far as possible every working man and woman in the nation safe and healthful working conditions and to preserve our human resources." This law took effect in April 1972.

The administration and enforcement of this Act are primarily vested in the Occupational Safety and Health Administration (OSHA) of the U.S. Department of Labor and an agency created for the purpose, the Occupational Safety and Health Review Commission. Research and related functions are carried out by the National Institute for Occupational Safety and Health, which was established within the U.S. Department of Health, Education and Welfare.

The Occupational Safety and Health Act has been the cause of some concern mainly among smaller industrial enterprises lest the provisions of the Act threaten their competitive position.

It is not generally understood that many of the standards incorporated in the Act were drawn up and used long before the Act was passed.

Major divisions of U.S. industry such as the maritime, construction, electrical, and chemical industries and many unions as well as industrial fire and accident insurance underwriters have in the past issued regulations and standards pertaining to safe practices, covering many aspects of industrial operations. These standards and regulations were observed on a voluntary basis by a large number of companies concerned about safe working conditions for their employees. Many of these same standards have been, at least partially, incorporated in OSHA Standards. This has sometimes resulted in overlapping, an example of which is found in the (slightly varying) specifications covering scaffolding found in the General Requirements as well as in standards for scaffolding as required in the sections covering the maritime and construction industries. It is to be expected that this overlapping will eventually be eliminated by the unification of all standards.

As to the perceived threat to a business' competitive position, two facts should be considered: First, most advanced industrial nations, especially those in northwestern Europe, have had similar legislation on their books for several years, and this legislation does not seem to have influenced their competitive position in any way. Second, many insurance companies that have had considerable experience in underwriting industrial hazards have statistically reached the conclusion that the practice of safety is good business and that it pays off in fewer lost hours resulting from accidents and illness caused by unhealthy working conditions, in better public and union relations, and in lower personnel turnover. These considerations should be ample justification of the Act to those who are not moved by its obvious and compelling humanitarian aspects.

OSHA Standards are many and cover a wide range of industries, but the guiding principles for establishing all these standards are the same, namely, a genuine interest in the health and safety of the worker and the application of common sense in accident prevention.

The field of mechanical power transmission is covered in part 1910 of OSHA Standards, 1910.219, entitled "Mechanical Power Transmission Apparatus," which is reprinted in its entirety at the end of this chapter.

Standard 1910.219 deals with flywheels, belts and pulleys, sprockets and chain, gearing, shafting, clutches, couplings and related parts, and so on.

The articles of this standard define the various hazardous conditions and configurations of power transmitting elements, and they indicate the different ways of guarding required. Also, construction and materials for guarding are described in some detail.

The following pages contain illustrations reprinted from the OSHA publication entitled *The Principles of Mechanical Guarding.* (OSHA 2057)

Figure 14.1 shows some typical hazards occurring in rotating power transmission elements.

Figure 14.2 deals with the so-called "in-running nip points" present in several kinds of power transmission element referred to in this standard.

Figures 14.3 through 14.18 show typical examples of guarding structures.

Figure 14.19 shows safe setscrew types and ditto shaft couplings.

Examples of hazards of
rotating elements

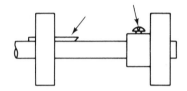

Rotating shaft and pulleys with projecting
key and set screw

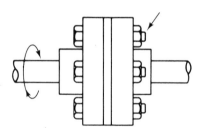

Rotating coupling with projecting bolt
heads

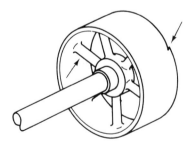

Rotating pulley with spokes and projecting burr on
face of pulley

**Figure 14.1**

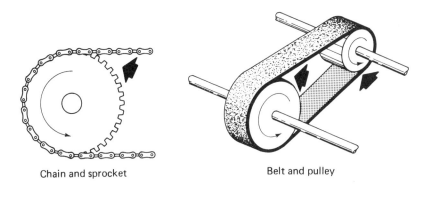

Chain and sprocket                    Belt and pulley

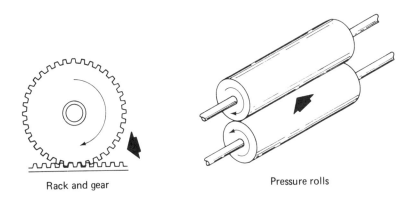

Rack and gear                         Pressure rolls

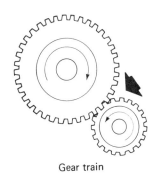

Gear train

**Figure 14.2**

Fly wheel guard — barrier for
flywheel and flywheel pit guard

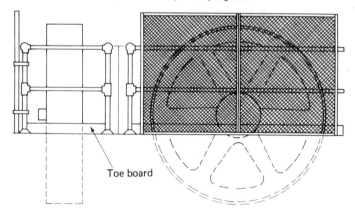

Toe board

**Figure 14.3**

Spur gear guards

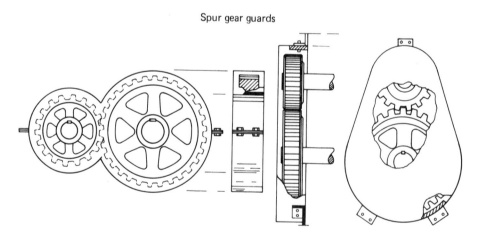

**Figure 14.4**

Pull cap with inclined belts

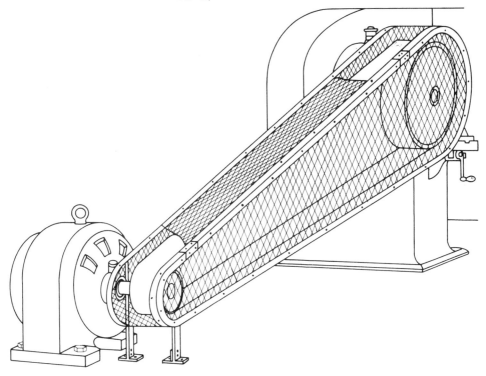

**Figure 14.5**

Overhead horizontal belt and pulley

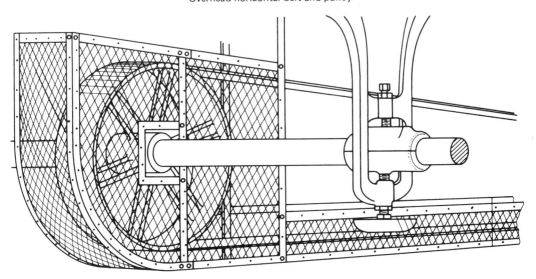

**Figure 14.6**

Pulleys and inclined belt

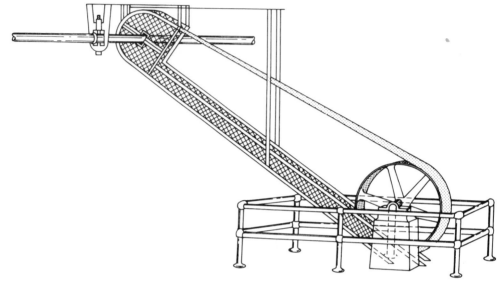

**Figure 14.7**

Fly wheel with horizontal belt

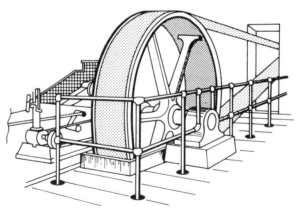

**Figure 14.8**

Horizontal shafting

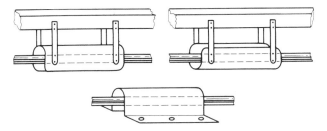

**Figure 14.9**

Horizontal shafting

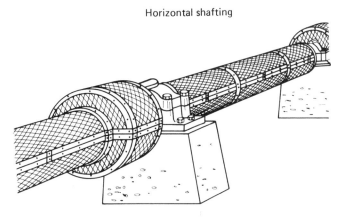

**Figure 14.10**

Horizontal shafting, belt and pulley

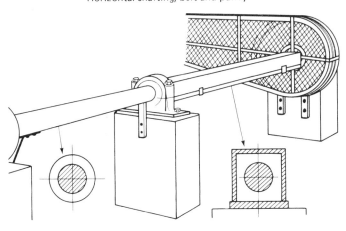

**Figure 14.11**

Vertical shafting

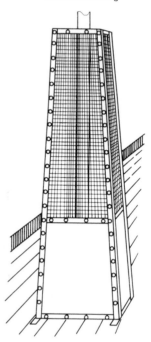

**Figure 14.12**

Sleeve for shaft end

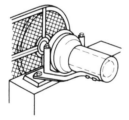

**Figure 14.13**

Coupling guard

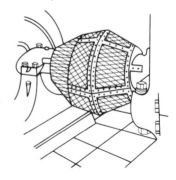

**Figure 14.14**

Bevel gear guards

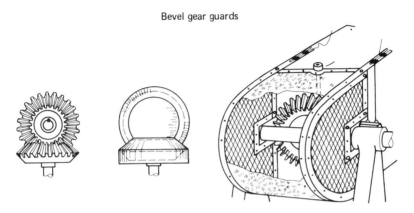

**Figure 14.15**

Elevated conveyor belt,
pulleys and tightener guard

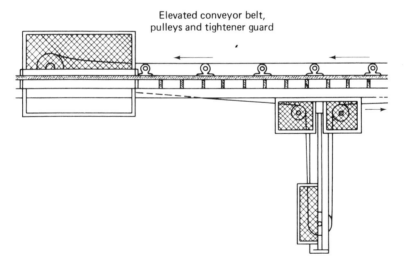

**Figure 14.16**

Calendar stock driveguard

Guarding in-running nip points by enclosure guards

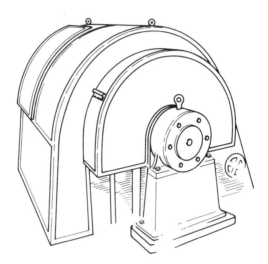

Figure 14.17

Belt and pulley guards

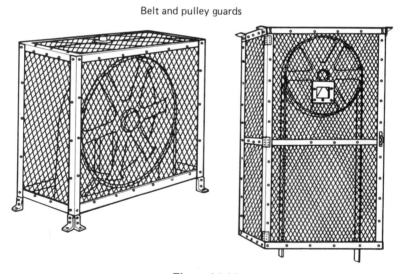

Figure 14.18

Safety shaft couplings

Elimination of rotating motion hazard by means other than guarding

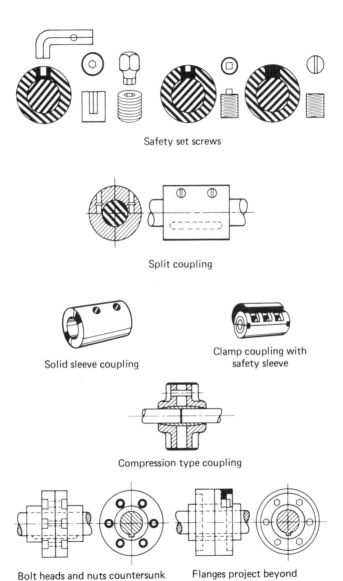

Safety set screws

Split coupling

Solid sleeve coupling

Clamp coupling with
safety sleeve

Compression type coupling

Bolt heads and nuts countersunk

Flanges project beyond
bolt heads and nuts

**Figure 14.19**

§1910.219 Mechanical power-transmission apparatus.

(a) *General requirements.* (1) This section covers all types and shapes of power-transmission belts, except the following when operating at two hundred and fifty (250) feet per minute or less: (i) Flat belts one (1) inch or less in width, (ii) flat belts two (2) inches or less in width which are free from metal lacings or fasteners, (iii) round belts one-half ($\frac{1}{2}$) inch or less in diameter; and (iv) single strand V-belts, the width of which is thirteen thirty-seconds ($\frac{13}{32}$) inch or less.

(2) Vertical and inclined belts (paragraphs (e) (3) and (4) of this section) if not more than two and one-half ($2\frac{1}{2}$) inches wide and running at a speed of less than one thousand (1,000) feet per minute, and if free from metal lacings or fastenings may be guarded with a nip-point belt and pulley guard.

(3) For the Textile Industry, because of the presence of excessive deposits of lint, which constitute a serious fire hazard, the sides and face sections only of nip-point belt and pulley guards are required, provided the guard shall extend at least six (6) inches beyond the rim of the pulley on the in-running and off-running sides of the belt and at least two (2) inches away from the rim and face of the pulley in all other directions.

(4) This section covers the principal features with which power transmission safeguards shall comply.

(b) *Prime-mover guards*—(1) *Flywheels.* Flywheels located so that any part is seven (7) feet or less above floor or platform shall be guarded in accordance with the requirements of this subparagraph:

(i) With an enclosure of sheet, perforated, or expanded metal, or woven wire;

(ii) With guard rails placed not less than fifteen (15) inches nor more than twenty (20) inches from rim. When flywheel extends into pit or is within 12 inches of floor, a standard toeboard shall also be provided;

(iii) When the upper rim of flywheel protrudes through a working floor, it shall be entirely enclosed or surrounded by a guardrail and toeboard.

(iv) For flywheels with smooth rims five (5) feet or less in diameter, where the preceding methods cannot be applied, the following may be used: A disk attached to the flywheel in such a manner as to cover the spokes of the wheel on the exposed side and present a smooth surface and edge, at the same time providing means for periodic inspection. An open space, not exceeding four (4) inches in width, may be left between the outside edge of the disk and the rim of the wheel if desired, to facilitate turning the wheel over. Where a disk is used, the keys or other dangerous projections not covered by disk shall be cut off or covered. This subdivision does not apply to flywheels with solid web centers.

(v) Adjustable guard to be used for starting engine or for running adjustment may be provided at the flywheel of gas or oil engines. A slot opening for jack bar will be permitted.

(vi) Wherever flywheels are above working areas, guards shall be installed having sufficient strength to hold the weight of the flywheel in the event of a shaft or wheel mounting failure.

(2) *Cranks and connecting rods.* Cranks and connecting rods, when exposed to contact shall be guarded in accordance with paragraphs (m) and (n) of this section, or by a guardrail as described in paragraph (o)(5) of this section.

(3) *Tail rods or extension piston rods.* Tail rods or extension piston rods shall be guarded in accordance with paragraphs (m) and (o) of this section, or by a guardrail on sides and end, with a clearance of not less than fifteen (15) nor more than twenty (20) inches when rod is fully extended.

(4) *Governor balls.* Governor balls six (6) feet or less from the floor or other working level, when exposed to contact, shall be provided with an enclosure extending to the top of the governor balls when at their highest position. The material used in the construc-

*Federal Register*, Vol. 37, No. 202 — Wednesday, October 18, 1972.

tion of this enclosure shall conform to paragraphs (m) and (o) of this section.

(c) *Shafting*—(1) *Installation.* (i) Each continuous line of shafting shall be secured in position against excessive endwise movement.

(ii) Inclined and vertical shafts, particularly inclined idler shafts, shall be securely held in position against endwise thrust.

(2) *Guarding horizontal shafting.* (i) All exposed parts of horizontal shafting seven (7) feet or less from floor or working platform, excepting runways used exclusively for oiling, or running adjustments, shall be protected by a stationary casing enclosing shafting completely or by a trough enclosing sides and top or sides and bottom of shafting as location requires.

(ii) Shafting under bench machines shall be enclosed by a stationary casing, or by a trough at sides and top or sides and bottom, as location requires. The sides of the trough shall come within at least six (6) inches of the underside of table, or if shafting is located near floor within six (6) inches of floor. In every case the sides of trough shall extend at least two (2) inches beyond the shafting or protuberance.

(3) *Guarding vertical and inclined shafting.* Vertical and inclined shafting seven (7) feet or less from floor or working platform, excepting maintenance runways, shall be enclosed with a stationary casing in accordance with requirements of paragraphs (m) and (o) of this section.

(4) *Projecting shaft ends.* (i) Projecting shaft ends shall present a smooth edge and end and shall not project more than one-half the diameter of the shaft unless guarded by nonrotating caps or safety sleeves.

(ii) Unused keyways shall be filled up or covered.

(5) *Power-transmission apparatus located in basements.* All mechanical power transmission apparatus located in basements, towers, and rooms used exclusively for power transmission equipment shall be guarded in accordance with this section, except that the requirements for safeguarding belts, pulleys, and shafting need not be complied with when the following requirements are met:

(i) The basement, tower, or room occupied by transmission equipment is locked against unauthorized entrance.

(ii) The vertical clearance in passageways between the floor and power transmission beams, ceiling, or any other objects, is not less than five feet six inches (5 ft. 6 in.).

(iii) The intensity of illumination conforms to the requirements of ANSI A11.1-1965 (R-1970).

(iv) The footing is dry, firm, and level.

(v) The route followed by the oiler is protected in such manner as to prevent accident.

(d) *Pulleys*—(1) *Guarding.* Pulleys, any parts of which are seven (7) feet or less from the floor or working platform, shall be guarded in accordance with the standards specified in paragraphs (m) and (o) of this section. Pulleys serving as balance wheels (e.g., punch presses) on which the point of contact between belt and pulley is more than six feet six inches (6 ft. 6 in.) from the floor or platform may be guarded with a disk covering the spokes.

(2) *Location of pulleys.* (i) Unless the distance to the nearest fixed pulley, clutch, or hanger exceeds the width of the belt used, a guide shall be provided to prevent the belt from leaving the pulley on the side where insufficient clearance exists.

(ii) Where there are overhanging pulleys on line, jack, or countershafts with no bearing between the pulley and the outer end of the shaft, a guide to prevent the belt from running off the pulley should be provided.

(3) *Broken pulleys.* Pulleys with cracks, or pieces broken out of rims, shall not be used.

(4) *Pulley speeds.* Pulleys intended to operate at rim speed in excess of manufacturers normal recommendations shall be specially designed and carefully balanced for the speed at which they are to operate.

(5) *Composition and wood pulleys.* Composition or laminated wood pulleys shall not be installed where they are subjected to influences detrimental to their structural composition.

(e) *Belt, rope, and chain drives*—(1) *Horizontal belts and ropes.* (i) Where both runs of horizontal belts are seven (7) feet or less from the floor level, the guard shall extend to at least fifteen (15) inches above the belt or to a standard height (see Table O-12), except that where both runs of a horizontal belt are 42 inches or less from the floor, the belt shall be fully enclosed in accordance with paragraphs (m) and (o) of this section.

(ii) In powerplants or power development rooms, a guardrail may be used in lieu of the guard required by subdivision (i) of this subparagraph.

(2) *Overhead horizontal belts.* (i) Overhead horizontal belts, with lower parts seven (7) feet or less from the floor or platform, shall be guarded on sides and bottom in accordance with paragraph (o)(3) of this section.

(ii) Horizontal overhead belts more than seven (7) feet above floor or platform shall be guarded for their entire length under the following conditions:

(*a*) If located over passageways or work places and traveling 1,800 feet or more per minute.

(*b*) If center to center distance between pulleys is ten (10) feet or more.

(*c*) If belt is eight (8) inches or more in width.

(iii) Where the upper and lower runs of horizontal belts are so located that passage of persons between them would be possible, the passage shall be either:

(*a*) Completely barred by a guardrail or other barrier in accordance with paragraphs (m) and (o) of this section; or

(*b*) Where passage is regarded as necessary, there shall be a platform over the lower run guarded on either side by a railing completely filled in with wire mesh or other filler, or by a solid barrier. The upper run shall be so guarded as to prevent contact therewith either by the worker or by objects carried by him. In powerplants only the lower run of the belt need be guarded.

(iv) Overhead chain and link belt drives are governed by the same rules as overhead horizontal belts and shall be guarded in the same manner as belts.

(v) American or Continuous System rope drives so located that the condition of the rope (particularly the splice) cannot be constantly and conveniently observed, shall be equipped with a telltale device (preferably electric-bell type) that will give warning when rope begins to fray.

(3) *Vertical and inclined belts.* (i) Vertical and inclined belts shall be enclosed by a guard conforming to standards in paragraphs (m) and (o) of this section.

(ii) All guards for inclined belts shall be arranged in such a manner that a minimum clearance of seven (7) feet is maintained between belt and floor at any point outside of guard.

(4) *Vertical belts.* Vertical belts running over a lower pulley more than seven (7) feet above floor or platform shall be guarded at the bottom in the same manner as horizontal overhead belts, if conditions are as stated in subparagraphs (2)(ii) (*a*) and (*c*) of this paragraph.

(5) *Cone-pulley belts.* (i) The cone belt and pulley shall be equipped with a belt shifter so constructed as to adequately guard the nip point of the belt and pulley. If the frame of the belt shifter does not adequately guard the nip point of the belt and pulley, the nip point shall be further protected by means of a vertical guard placed in front of the pulley and extending at least to the top of the largest step of the cone.

(ii) If the belt is of the endless type or laced with rawhide laces, and a belt shifter is not desired, the belt will be considered guarded if the nip point of the belt and pulley is protected by a nip point guard located in front of the cone extending at least to the top of the largest step of the cone, and formed to show the contour of the cone in order to give the nip point of the belt and pulley the maximum protection.

(iii) If the cone is located less than 3 feet from the floor or working platform, the cone pulley and belt shall be guarded to a height of 3 feet regardless of whether the belt is endless or laced with rawhide.

(6) *Belt tighteners.* (i) Suspended counterbalanced tighteners and all parts thereof shall be of substantial construction and securely fastened; the bearings shall be securely capped. Means must be provided to prevent tightener from falling, in case the belt breaks.

(ii) Where suspended counterweights are used and not guarded by location, they shall be so encased as to prevent accident.

(f) *Gears, sprockets, and chains—(1) Gears.* Gears shall be guarded in accordance with one of the following methods:

(i) By a complete enclosure; or

(ii) By a standard guard as described in paragraph (o) of this section, at least seven (7) feet high extending six (6) inches above the mesh point of the gears; or

(iii) By a band guard covering the face of

gear and having flanges extended inward beyond the root of the teeth in the exposed side or sides. Where any portion of the train of gears guarded by a band guard is less than six (6) feet from the floor a disk guard or a complete enclosure to the height of six (6) feet shall be required.

(2) *Hand-operated gears.* Subparagraph (1) of this paragraph does not apply to hand-operated gears used only to adjust machine parts and which to not continue to move after hand power is removed. However, the guarding of these gears is highly recommended.

(3) *Sprockets and chains.* All sprocket wheels and chains shall be enclosed unless they are more than seven (7) feet above the floor or platform. Where the chain extends over other machine or working areas, protection against falling shall be provided. This subparagraph does not apply to manually operated sprockets.

(4) *Openings for oiling.* When frequent oiling must be done, openings with hinged or sliding self-closing covers shall be provided. All points not readily accessible shall have oil feed tubes if lubricant is to be added while machinery in in operation.

(g) *Guarding friction drives.* The driving point of all friction drives when exposed to contact shall be guarded, all spoke friction drives and all web friction drives with holes in the web shall be entirely enclosed, and all projecting bolts on friction drives where there may be contact shall be guarded.

(h) *Keys, setscrews, and other projections.* (1) All projecting keys, setscrews, and other projections in revolving parts shall be removed or made flush or guarded by metal cover. This subparagraph does not apply to keys or setscrews within gear or sprocket casings or other enclosures, nor to keys, setscrews, or oilcups in hubs of pulleys less than twenty (20) inches in diameter when they are within the plane of the rim of the pully.

(2) It is recommended, however, that no projecting setscrews or oilcups be used in any revolving pulley or part of machinery.

(i) *Collars and couplings*—(1) *Collars.* All revolving collars, including split collars, shall be cylindrical, and screws or bolts used in collars shall not project beyond the largest periphery of the collar.

(2) *Couplings.* Shaft couplings shall be so constructed as to present no hazard from bolts, nuts, setscrews, or revolving surfaces. Bolts, nuts, and setscrews will, however, be permitted where they are covered with safety sleeves or where they are used parallel with the shafting and are countersunk or else do not extend beyond the flange of the coupling.

(j) *Bearings and facilities for oiling.* Self lubricating bearings are recommended and all drip cups and pans shall be securely fastened.

(k) *Guarding of clutches, cutoff couplings, and clutch pulleys*—(1) *Guards.* Clutches, cutoff couplings, or clutch pulleys having projecting parts, where such clutches are located seven (7) feet or less above the floor or working platform, shall be enclosed by a stationary guard constructed in accordance with this section. A "U" type guard is permissible.

(2) *Engine rooms.* In engine rooms a guardrail, preferably with toeboard, may be used instead of the guard required by subparagraph (1) of this paragraph, provided such a room is occupied only by engine room attendants.

(3) *Bearings.* A bearing support immediately adjacent to a friction clutch or cutoff coupling shall have self-lubricating bearings requiring attention at infrequent intervals.

(l) *Belt shifters, clutches, shippers, poles, perches, and fasteners*—(1) *Belt shifters.* (i) Tight and loose pulleys on all new installations made on or after August 31, 1971, shall be equipped with mechanical means to prevent belt from creeping from loose to tight pulley. It is recommended that old installations be changed to conform to this rule.

(ii) Belt shifter and clutch handles shall be rounded and be located as far as possible from danger of accidental contact, but within easy reach of the operator. Where belt shifters are not directly located over a machine or bench, the handles shall be cut off six feet six inches (6 ft. 6 in.) above floor level.

(iii) All belt and clutch shifters of the same type in each shop should move in the same direction to stop machines, i.e., either all right or all left. This does not apply to friction clutch on countershaft carrying two clutch pulleys with open and crossed belts, respectively. In this case the shifter handle has three positions and the machine is at a standstill when clutch handle is in the neutral or center position.

(2) *Belt shippers and shipper poles.* The use of belt poles as substitutes for mechanical shifters is not recommended. Where necessity compels their use, they shall be of sufficient size to enable workmen to grasp them securely. (A two-inch (2 in.) diameter or 1½ by 2 inches cross-section is suggested.) Poles shall be smooth and preferably of straight grain hardwood, such as ash or hickory. The edges of rectangular poles should be rounded. Poles should extend from the top of the pulley to within about forty (40) inches of floor or working platform.

(3) *Belt perches.* Where loose pulleys or idlers are not practicable, belt perches in form of brackets, rollers, etc., shall be used to keep idle belts away from the shafts. Perches should be substantial and designed for the safe shifting of belts.

(4) *Belt fasteners.* Belts which of necessity must be shifted by hand and belts within seven (7) feet of the floor or working platform which are not guarded in accordance with this section shall not be fastened with metal in any case, nor with any other fastening which by construction or wear will constitute an accident hazard.

(m) *Standard guards—general requirements—*(1) *Materials.* (i) Standard conditions shall be secured by the use of the following materials. Expanded metal, perforated or solid sheet metal, wire mesh on a frame of angle iron, or iron pipe securely fastened to floor or to frame of machine.

(ii) All metal should be free from burrs and sharp edges.

(iii) Wire mesh should be of the type in which the wires are securely fastened at every cross point either by welding, soldering, or galvanizing, except in case of diamond or square wire mesh made of No. 14 gage wire, ¾-inch mesh or heavier.

(2) *Methods of manufacture.* (i) Expanded metal, sheet or perforated metal, and wire mesh shall be securely fastened to frame by one of the following methods:

(*a*) With rivets or bolts spaced not more than five (5) inches center to center. In case of expanded metal or wire mesh, metal strips or clips shall be used to form a washer for rivets or bolts.

(*b*) by welding to frame every four (4) inches.

(*c*) By weaving through channel or angle frame, or if No. 14 gage ¾-inch mesh or heavier is used by bending entirely around rod frames.

(*d*) Where openings in pipe railing are to be filled in with expanded metal, wire mesh or sheet metal, the filler material shall be made into panels with rolled edges or bound with "V" or "U" edging of No. 24 gage or heavier sheet metal fastened to the panels with bolts or rivets spaced not more than five (5) inches center to center. The bound panels shall be fastened to the railing by sheet-metal clips spaced nor more than five (5) inches center to center.

(*e*) Diamond or square mesh made of crimped wire fastened into channels, angle or round-iron frames, may also be used as a filler in guards. Size of mesh shall correspond to Table O-12.

(ii) Where the design of guards requires filler material of greater area than 12 square feet, additional frame members shall be provided to maintain panel area within this limit.

(iii) All joints of framework shall be made equivalent in strength to the material of the frame.

(n) *Disk, shield, and "U" guards—*(1) *Disk guards.* A disk guard shall consist of a sheet-metal disk not less than No. 22 gage fastened by "U" bolts or rivets to spokes of pulleys, flywheels, or gears. Where possibility of contact with sharp edges of the disk exists, the edge shall be rolled or wired. In all cases the nuts shall be provided with locknuts which shall be placed on the unexposed side of the wheel.

(2) *Shield guards.* (i) A shield guard shall consist of a frame filled in with wire mesh, expanded, perforated, or solid sheet metal.

(ii) If area of shield does not exceed six (6) square feet the wire mesh or expanded metal may be fastened in a framework of $^3/_8$-inch solid rod, $^3/_4$-inch by $^3/_4$-inch by $^1/_8$-inch angle iron or metal construction of equivalent strength. Metal shields may have edges entirely rolled around a $^3/_8$-inch solid iron rod.

(3) *"U" guards.* A "U" guard consisting of a flat surface with edge members shall be designed to cover the under surface and lower edge of a belt, multiple chain, or rope drive. It shall be constructed of materials specified in Table O-12, and shall conform to the requirements of paragraphs (o)(3) and (4) of this sec-

## TABLE O-12

*Table of Standard Materials and Dimensions*

| Material | Clearance from moving part at all points | Largest mesh or opening allowable | Minimum gauge (U.S. Standard) or thickness | Minimum height of guard from floor or platform level |
|---|---|---|---|---|
| | *Inches* | *Inches* | | *Feet* |
| Woven wire | Under 2 | $^3/_8$ | No. 16 | 7 |
| | 2-4 | $^1/_2$ | No. 16 | 7 |
| | Under 4 | $^1/_2$ | No. 16 | 7 |
| | 4-15 | 2 | No. 12 | 7 |
| Expanded metal | Under 4 | $^1/_2$ | No. 18 | 7 |
| | 4-15 | 2 | No. 13 | 7 |
| Perforated metal | Under 4 | $^1/_2$ | No. 20 | 7 |
| | 4-15 | 2 | No. 14 | 7 |
| Sheet metal | Under 4 | | No. 22 | 7 |
| | 4-15 | | No. 22 | 7 |
| Wood or metal strip crossed | Under 4 | $^3/_8$ | Wood $^3/_4$ Metal No. 16 | 7 |
| | 4-15 | 2 | Wood $^3/_4$ Metal No. 16 | 7 |
| Wood or metal strip not crossed | Under 4 | $^1/_2$ width | Wood $^3/_4$ Metal No. 16 | 7 |
| | 4-15 | 1 width | Wood $^3/_4$ Metal No. 16 | 7 |
| Standard rail | Min. 15 | | | |
| | Max. 20 | | | |

tion. Edges shall be smooth and, if size of guard requires, these edges shall be reinforced by rolling, wiring, or by binding with angle or flat iron.

(o) *Approved materials*—(1) *Minimum requirements.* The materials and dimensions specified in this paragraph shall apply to all guards, except horizontal overhead belts, rope, cable, or chain guards more than seven (7) feet above floor, or platform. (For the latter, see Table O-13.)

(i) Minimum dimensions of materials for the framework of all guards, except as noted in subdivision (i)(c) shall be angle iron 1 inch by 1 inch by $^1/_8$ inch, metal pipe of $^3/_4$-inch inside diameter or metal construction of equivalent strength.

(a) All guards shall be rigidly braced every three (3) feet or fractional part of their height to some fixed part of machinery or building structure. Where guard is exposed to contact with moving equipment additional strength may be necessary.

(b) The framework for all guards fastened to floor or working platform and without other support or bracing shall consist of 1½-inch by $^1/_8$-inch angle iron, metal pipe of 1½-inch inside diameter, or metal construction of equivalent strength. All rectangular guards shall have at least four upright frame members each of which shall be carried to the floor and be securely fastened thereto. Cylindrical guards shall have at least three supporting members carried to floor.

(c) Guards thirty (30) inches or less in height and with a total surface area not in excess of ten (10) square feet may have a framework of $^3/_8$-inch solid rod, $^3/_4$-inch by $^3/_4$-inch by $^1/_8$-inch angle, or metal construction of equivalent strength. The filling material shall correspond to the requirements of Table O-12.

(ii) The specifications given in Table O-12 and subdivision (i) of this subparagraph are minimum requirements; where guards are exposed to unusual wear, deterioration or impact, heavier material and construction should be used to protect amply against the specific hazards involved.

(2) *Wood guards.* (i) Wood guards may

## TABLE O-13

### Horizontal Overhead Belts, Ropes, and Chains 7 Feet or More Above Floor or Platform

| | Width | | | Material |
|---|---|---|---|---|
| | From 10" to 14" inclusive | Over 14" to 24" inclusive | Over 24" | |
| **MEMBERS** | | | | |
| Framework | $1^1/_2$" × $1^1/_2$" × $^1/_4$" | 2" × 2" × $^5/_{16}$" | 3" × 3" × $^3/_8$" | Angle iron. |
| Filler (belt guards) | $1^1/_2$" × $^8/_{16}$" | 2" × $^3/_{16}$" | 2" × $^5/_{16}$" | Flat iron. |
| Filler and vertical side member. | No. 20 A.W.G. | No. 18 A.W.G. | No. A.W.G. | Solid sheet metal. |
| Filler supports | 2" × $^5/_{16}$" flat iron | 2" × $^3/_8$" flat iron | $2^1/_2$" × $2^1/_2$" × $^1/_4$" angle | Flat and angle. |
| Guard supports | 2" × $^5/_{16}$" | 2" × $^3/_8$" | $2^1/_2$" × $^3/_8$" | Flat iron. |
| **FASTENINGS** | | | | |
| Filler supports to framework | (2) $^5/_{16}$" | (2) $^3/_8$" | (3) $^1/_2$" | Rivets. |
| Filler flats to supports (belt guards). | (1) $^5/_{16}$" | (1) $^5/_{16}$" | (2) $^3/_8$" | Flush rivets. |
| Filler to frame and supports (chain guards). | $^3/_{16}$" rivets spaced | 8" centers on sides and 4" centers on bottom. | . . . . . . . . . . . . . | |
| Guard supports to framework. | (2) $^3/_8$" | (2) $^7/_{16}$" | (2) $^5/_8$" | Rivets or bolts. |
| Guard and supports to overhead ceiling. | $^1/_4$" × $3^1/_2$" lag screws or $^1/_2$" bolts. | $^5/_8$" × 4" lag screws or $^5/_8$" bolts. | $^3/_4$" × 6" lag screws or $^3/_4$" bolts. | Lag screws or bolts. |
| **DETAILS – SPACING, ETC.** | | | | |
| Width of guards | One-quarter wider than belt, rope, or chain drive. | . . . . . . . . . . . . . . | | |
| Spacing between filler supports. | 20" C. to C. | 16" C. To C. | 16" C. to C. | |
| Spacing between filler flats (belt guards). | 2" apart | $2^1/_2$" apart | 3" apart | |
| Spacing between guard supports. | 36" C. to C. | 36" C. to C. | 36" C. to C. | |
| **OTHER BELT GUARD FILLING PERMITTED** | | | | |
| Sheet metal fastened as in chain guards. | No. 20 A.W.G. | No. 18 A.W.G. | No. 18 A.W.G. | Solid or perforated. |
| Woven wire, 2" mesh | No. 12 A.W.G. | No. 10 A.W.G. | No. 8 A.W.G. | |
| **CLEARANCE FROM OUTSIDE OF BELT, ROPE, OR CHAIN DRIVE TO GUARD** | | | | |
| Distance center to center of shafts. | Up to 15' inclusive | Over 15' to 25' inclusive. | Over 25' to 40' inclusive. | Over 40'. |
| Clearance from belt, or chain to guard. | 6" | 10" | 15" | 20". |

be used in the woodworking and chemical industries, in industries where the presence of fumes or where manufacturing conditions would cause the rapid deterioration of metal guards; also in construction work and in locations outdoors where extreme cold or extreme heat make metal guards and railings undesirable. In all other industries, wood guards shall not be used.

(ii)(*a*) Wood shall be sound, tough, and free from any loose knots.

(*b*) Guards shall be made of planed lumber not less than one (1) inch rough board measure, and edges and corners rounded off.

(*c*) Wood guards shall be securely fastened together with wood screws, hardwood dowel pins, bolts, or rivets.

(*d*) While no definite dimensions are given under this heading for framework or filler materials, wood guards shall be equal in strength and rigidity to metal guards specified in subparagraphs (1)(i) and (ii) of this paragraph and Table O-12.

(*e*) For construction of standard wood railing, see subparagraph (5) of this paragraph.

(3) *Guards for horizontal overhead belts.* (i) Guards for horizontal overhead belts shall run the entire length of the belt and follow the line of the pulley to the ceiling or be carried to the nearest wall, thus enclosing the belt effectively. Where belts are so located as to make it impracticable to carry the guard to wall or ceiling, construction of guard shall be such as to enclose completely the top and bottom runs of belt and the face of pulleys.

(ii) The guard and all its supporting members shall be securely fastened to wall or ceiling by gimlet-point lag screws or through bolts. In case of masonry construction, expansion bolts shall be used. The use of bolts placed horizontally through floor beams or ceiling rafters is recommended.

(iii) Suitable reinforcement shall be provided for the ceiling rafters or overhead floor beams, where such is necessary, to sustain safely the weight and stress likely to be imposed by the guard. The interior surface of all guards, by which is meant the surface of the guard with which a belt will come in contact, shall be smooth and free from all projections of any character, except where construction demands it; protruding shallow roundhead rivets may be used. Overhead belt guards shall

be at least one-quarter wider than the belt which they protect, except that this clearance need not in any case exceed six (6) inches on each side. Overhead rope drive and block and roller-chain-drive guards shall be not less than six (6) inches wider than the drive on each side. In overhead silent chain-drive guards where the chain is held from lateral displacement on the sprockets, the side clearances required on drives of twenty (20) inch centers or under shall be not less than one-fourth inch from the nearest moving chain part, and on drives of over twenty (20) inch centers a minimum of one-half inch from the nearest moving chain part.

(iv) Table O-13 gives the sizes of materials to be used and the general construction specifications of guards for belts ten (10) inches or more in width. No material for overhead belt guards should be smaller than that specified in Table O-13 for belts ten (10) to fourteen (14) inches wide, even if the overhead belt is less than ten (10) inches in width. However, No. 20 gage sheet metal may be used as a filler on guards for belts less than ten (10) inches wide. Expanded metal, because of the sharp edges, should not be used as a filler in horizontal belt guards.

(v) For clearance between guards and belts, ropes or chains of various center to center dimensions between the shafts, see bottom of Table O-13.

(4) *Guards for horizontal overhead rope and chain drives.* Overhead-rope and chain-drive guard construction shall conform to the rules for overhead-belt guard construction of similar width, except that the filler material shall be of the solid type as shown in Table O-13, unless the fire hazard demands the use of open construction. A side guard member of the same solid filling material should be carried up in a vertical position two (2) inches above the level of the lower run of the rope or chain drive and two (2) inches within the periphery of the pulleys which the guard encloses thus forming a trough. These side filler members should be reinforced on the edges with 1½-inch by ¼-inch flat steel, riveted to the filling material at not greater than eight (8) inch centers; the reinforcing strip should be fastened or bolted to all guard supporting members with at least one $3/8$-inch rivet or bolt at each intersection, and the ends should

be secured to the ceiling with lag screws or bolts. The filling material shall be fastened to the framework of the guard and the filler supports by $^3/_{16}$-inch rivets spaced on 4-inch centers. The width of the multiple drive shall be determined by measuring the distance from the outside of the first to the outside of the last rope or chain in the group accommodated by the pulley.

(5) *Guardrails and toeboards.* (i) Guardrail shall be forty-two (42) inches in height, with midrail between top rail and floor.

(ii) Posts shall be not more than eight (8) feet apart; they are to be permanent and substantial, smooth, and free from protruding nails, bolts, and splinters. If made of pipe, the post shall be one and one-fourth (1¼) inches inside diameter, or larger. If made of metal shapes or bars, their section shall be equal in strength to that of one and one-half (1½) by one and one-half (1½) by three-sixteenths ($^3/_{16}$) inch angle iron. If made of wood, the posts shall be two by four (2 × 4) inches or larger. The upper rail shall be two by four (2 × 4) inches, or two one by four (1 × 4) strips, one at the top and one at the side of posts. The midrail may be one by four (1 × 4) inches or more. The rails (metal shapes, metal bars, or wood), should be on that side of the posts which gives the best protection and support. Where panels are fitted with expanded metal or wire mesh as noted in Table O-12 the middle rails may be omitted. Where guard is exposed to contact with moving equipment, additional strength may be necessary.

(iii) Toeboards shall be four (4) inches or more in height, of wood, metal, or of metal grill not exceeding one (1) inch mesh. Toeboards at flywheel pits should preferably be placed as close to edge of the pit as possible.

(p) *Care of equipment*—(1) *General.* All power-transmission equipment shall be inspected at intervals not exceeding 60 days and be kept in good working condition at all times.

(2) *Shafting.* (i) Shafting shall be kept in alignment, free from rust and excess oil or grease.

(ii) Where explosives, explosive dusts, flammable vapors or flammable liquids exist, the hazard of static sparks from shafting shall be carefully considered.

(3) *Bearings.* Bearings shall be kept in alignment and properly adjusted.

(4) *Hangers.* Hangers shall be inspected to make certain that all supporting bolts and screws are tight and that supports of hanger boxes are adjusted properly.

(5) *Pulleys.* (i) Pulleys shall be kept in proper alignment to prevent belts from running off.

(ii) One or both pulleys carrying a non-shifting belt should have crowned faces.

(iii) Cast-iron pulleys should be tested frequently with a hammer to disclose cracks in rim or spokes. It should be borne in mind that the sound is usually the pulley.

(iv) Split pulleys should be inspected to ascertain if all bolts holding together the sections of the pulley are tight.

(6) *Care of belts.* (i) Quarter-twist belts when installed without an idler can be used on drives running in one direction only. They will run off a pulley when direction of motion is reversed.

(ii) Inspection shall be made of belts, lacings, and fasteners and such equipment kept in good repair.

(iii) Where possible, dressing should not be applied when belt or rope is in motion; but, if this is necessary, it should be applied where belts or rope leave pulley, not where they approach. The same precautions apply to lubricating chains. In the case of V-belts, belt dressing is neither necessary nor advisable.

(7) *Lubrication.* The regular oilers shall wear tight-fitting clothing and should use cans with long spouts to keep their hands out of danger. Machinery shall be oiled when not in motion, wherever possible.

§**1910.220 Effective dates.**

(a) The provisions of this Subpart O shall become effective on August 27, 1971, except as provided in the remaining paragraphs of this section.

(b) The following provisions shall become effective on February 15, 1972:
§1910.212(a).
§1910.213(a), (b), (c), (d), (e), (f), (g), (h), (i), (j), (k), (l), (m), (n), (o), (p), (q), and (r).
§1910.214(a), (b), (c), (d), (e), (f), (g), (h), (i), (j), (k), (l), (m), (n), (o), (p), (q), (r), (s), (t), (u), and (v).
§1910.215(a) and (b).
§1910.216(a), (b), (c), and (f).

§1910.217(a), (b), (c), and (d).

§1910.218(a), (b), (d), (e), (g), and (j).

§1910.219(b), (c), (d), (e), (f), (g), (h), (i), and (k).

(c) Notwithstanding anything in paragraph (a), (b), or (d) of this section, any provision in any other section of this subpart which contains in itself a specific effective date or time limitation shall become effective on such date or shall apply in accordance with such limitation.

(d) Notwithstanding anything in paragraph (a) or (b) of this section, if any standard in 41 CFR Part 50-204, other than a national consensus standard incorporated by reference in §50-204.2(a)(1), is or becomes applicable at any time to any employment and place of employment, by virtue of the Walsh-Healey Public Contracts Act, or the Service Contract Act of 1965, or the National Foundation on Arts and Humanities Act of 1965, any corresponding established Federal standard in this Subpart O which is derived from 41 CFR Part 50-204 shall also become effective, and shall be applicable to such employment and place of employment, on the same date.

[36 F.R. 15106, Aug. 13, 1971]

# APPENDIX

*Metric conversion factors*

1 ft-lb (foot-pound) = .138 joule (*J*)
1 kgF (kilogram force) = 2.2 lbF (pound force)
1 N (Newton) = .225 lbF
1 Nm (Newton-meter) = .737 ft-lb
1 hp (horsepower) = 33,000 ft-lb/min = 550 ft-lb/s = 746 watts (*W*)
1 kW (kilowatt) = 1.34 hp

# INDEX